MICROELECTRONICS EDUCATION

MICROELECTRONICS EDUCATION

Proceedings of the 2nd European Workshop held in Noordwijkerhout, The Netherlands, 14-15 May 1998

Edited by

Ton J. Mouthaan

and

Cora Salm

MESA Research Institute,
University of Twente,
Enschede, The Netherlands

SPRINGER-SCIENCE+BUSINESS MEDIA, B.V.

A C.I.P. Catalogue record for this book is available from the Library of Congress.

ISBN 978-0-7923-5107-8 ISBN 978-94-011-5110-8 (eBook)
DOI 10.1007/978-94-011-5110-8

Printed on acid-free paper

Foreword

Dear participant in the second European Workshop on Microelectronics Education,

It is a pleasure to present you the Proceedings of the Second European Workshop on Microelectronics Education and to welcome you at the Workshop.

The Organising Committee is very pleased that it has found several key persons, with highly appreciated levels of knowledge and expertise, willing to present Invited Contributions to this Workshop. We have striven for an interesting spread over important areas like the expected demands for educated engineers in the wide field of Microelectronics, and Microsystems, in European industry (and beyond!) and innovations in method and focus of our educational programmes.

This is the second European Workshop in this area; the first one was held in Grenoble in France in the spring of 1996. It was the initiative of Georges Kamarinos, Nadine Guillemot and Bernard Courtois to organise this Workshop because they felt that Microelectronics was 'at a turning point' to become the core of the largest industry in the world and that this warranted a serious (re-)consideration of our educational imperatives. It is now two years since and their feeling has become reality: nobody doubts that by the year 2000 the microelectronics industry will be the largest industrial sector. It is also obvious that because of that and because of the predicted shortfall of educated engineers we must continuously reconsider the quality of our educational approach.

There are two main questions here: are we teaching the right things and are we teaching them the right way? Nobody can claim to have an answer to these questions but it is through Workshops like this one that we can open our minds to people who have formulated answers from their points of view. May be some consensus may arise from this and almost certainly this Workshop will give you new ideas and inspiration.

The format of the Workshop emphasises Work and Shop. There are a very limited number of oral presentations: only those contributions that will interest a large number of people next to the invited papers are presented orally in extended form. Each session is followed by a panel discussion instead of discussions after each paper; this is the 'Work' part. The 'Shop' part is formed by the demonstrations/poster presentations. Here we have sessions lasting 2 hours or more in which a maximum of 8 short oral presentations are given followed by demonstrations and a more personal interaction.

People can walk around and 'Shop' to fill their minds with new ideas and meet the people behind the ideas.

Our close interaction with the American counterpart of this Workshop, the MicroSystems Education conference, is also new. We are pleased to welcome representatives of this conference in our midst and it is the intention to closely relate both events in such a way that we have a yearly meeting, alternating between the US and Europe. From the programme you can see this international orientation with invited speakers from the US, Canada, Japan and Australia. We have received abstracts from 30 different countries, so the Workshop truly is an International Workshop!

It is the intention of the organisers that a Workshop like this becomes a forum of which everyone feels it is the natural meeting place for people in our profession; if you are serious about your job, you should be there! It is not a scientific conference pur sang; we do not have a best paper award since the mere effort of everyone to reflect and be innovative is best practice in itself.

We hope you enjoy the Workshop.

The Organising Committee,

Ton Mouthaan
MESA Institute, Univ. of Twente

Nadine Guillemot
Centre Interuniversitaire de MicroElectronique

Jan Fluitman
MESA Institute, Univ. of Twente

Thursday 14 May

	8.45	9.15	9.45	10.15	10.45	11.15	11.35	12.00	14.00	14.45	15.00	15.15	15.30	16.00	16.15	16.45	18.45	20.30
Rotonde	Opening	**Session A** Industrial Outlook A1	A2	Coffee Break	**Session A** Continued A3	A4	Panel	Lunch Posters Demos	**Session B** Emerging Fields Invited B1	B2	B3	B4	Coffee Break	**B:** Cont`d B5	Panel			Conference Dinner (Leeuwenhorst) 20.30
Room B										**Sess. C** New Concepts C1	C2	C3	Coffee Break	**C:** Cont`d C4	Panel			
Foyer								**Session P1:** Industrial Projects / New Concepts	14.45 15.00 15.15 15.30							**P3:** Multi Media		
Lounge								**Session P2:** Emerging Fields / Technology								**P4:** Design Innovat.		

Friday 15 May

Room	8.30–10.00	10.00–10.30	10.30–11.30	11.30–13.30	13.30–15.45	15.45–16.00
Rotonde	**Session D** International Outlook: D1, D2, D3	Coffee Break	**Session D** Continued: D4, Panel	Lunch Posters Demos	**Session E** Multi Media: Invited E1, E2, E3, E4, E5, Panel	Closing
Room B				Lunch Posters Demos	**Session F** Design Innovations: Invited F1, F2, F3, F4, F5, Panel	Closing
Foyer				**Session P5** Multi Media		
Lounge				**Session P6** Design Innovations		

Time marks: 8.30, 9.00, 9.30, 10.00, 10.30, 11.00, 11.30, 13.30, 14.15, 14.30, 14.45, 15.00, 15.15, 15.45, 16.00

CONTENTS

Session C: New Concepts in teaching

Session P3: Multimedia in Microelectronics Education

Session A

INDUSTRIAL OUTLOOK

THE VIEW OF SIEMENS ON TRAINING, COLLABORATIVE PROJECTS, CONTINUING EDUCATION

H. SCHMÖKEL, T. FALTER, W. BEINVOGL
Siemens AG HL MP PT
Munich, Germany

1 Introduction

The continuously increasing speed and sequence in the development of new products in the wide field of microelectronics demands an extraordinary growth of the related engineering teams world wide in development and production. One of the most challenging tasks of a company is the recruiting and the training of new employees on the job which is a costly and time consuming task. The optimal employee with a high grade of special knowledge needed for a specific job cannot be expected from the university. But it is the task of the universities to offer general engineering knowledge, semiconductor basics, and soft skills to the students in order to give them the possibility to form their own personality. The specific requirements of Siemens semiconductor development division towards a potential candidate are shown in figure 1..

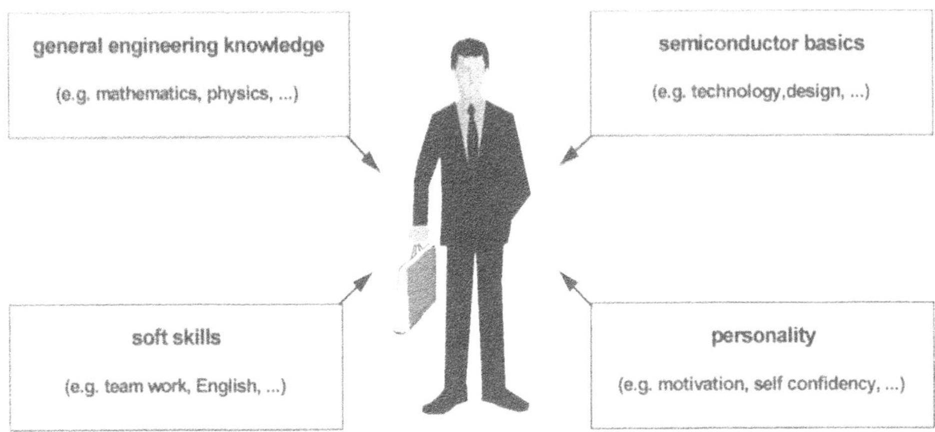

Figure 1: Requirements towards a potential candidate

T.J. Mouthaan and C. Salm (eds.), Microelectronics Education, 3-5.

In this paper the view of Siemens of proper university education, collaborative projects with universities and industrial research, and continuing education within or outside the company will be discussed. The goal must be to improve the already existing and well established education process in order to provide the industry with well prepared candidates. The training before the job in industry has to start at the university, the "initial" training is done in the first job in industry, and the continuing education of the employee will be done lifelong in the job.

2 Basic education at universities

The major task of the universities is teaching the technical background to give the student the general engineering knowledge for their job. The basics of the commonly used information for engineers are mathematics incl. statistics, physics, electrical engineering incl. high frequency technique, and electrical measurement techniques. Other subjects like mechanics, chemistry, and construction should be offered more as an option for specific knowledge. Basics in programming skills and network know how are obligatory for every engineer.
Semiconductor basics like process technology, the principles of design methodologies, digital circuitry have to be up to date but not more than 5 years old. In any chase the emphasis really is not only to learn but to understand the basics and principles in the best case by doing practical work and getting a feeling for the topics to solve real problems. A more detailed specialisation is not required due to the accelerated ageing of specific knowledge.
The following soft skills like the know how to learn that means how to get the necessary information within a very short time and how to use/transfer it (knowledge management) to solve technical problems is the main issue which should be trained by the universities. Because of the complexity of the technical issues problems can only be covered in teams. For this the basics of communication and presentation techniques have to be trained with practical work. In addition international experience and knowledge in foreign languages (English language is obligatory) is highly appreciated. This can be achieved by international study exchange programs at the universities. Candidates who have done a part of their study in a foreign country have improved language knowledge but also basic international experience.

3 Training on the job and continuing education

Starting a new job does not mean stop learning. In the first weeks new employees have to be prepared to the requirements of the new job. To become familiar with company specific equipment and software tools (e.g. testers, CAD-tools and test programs) training courses have to be attended.
Although it was pointed out in item 2 that soft skills are absolutely necessary, our experience shows that many candidates have quiet some deficiencies in that area. This

means that a company has to train the employees by specific courses like problem solving and decision making, communication training, intercultural training, and others. Continuing education is necessary due to new equipment, software tools, technologies or technical challenges during the whole working life of an engineer. This is caused by the very fast development of the technology and the increasing complexity of the knowledge also in hard- and software.
Continuing technical education in co operation with universities is appreciated. Additional courses have to be done due to job rotations or the increase of responsibility in the company e.g. project leader seminars, management support systems, ...

4 Collaborative projects

To get the optimum contact between university and industry with the highest benefits for both sides a company will prefer the close cooperation with dedicated universities. It is on the side of the industry to define together with the university more specific research projects which should be applicable in the near or medium scale future.
The gaol must be to get a close, bilateral information exchange. For this it is helpful to arrange frequent presentations and workshops which have to be driven by both sides. Besides this the master or Ph. D. theses are done either in the company or by request of the company at the university.

5 Summary

The overall present situation shows a lot of room for improvements. Due to the changing situation and the acceleration of knowledge increase and complexity it is recommended by Siemens semiconductor division to provide the students with up to date information. The goal must be to keep the contents up to date, to improve the methodology, and to focus also on soft skills. In any case the duration of a study must not be increased by these actions.
It is on all involved parties, the universities, the companies, and also on the students to drive the improvement actively. Only an internationally competitive university, an aggressive high tech company, and a student personality based on self confidence and flexibility fit together to a successful whole.

AN SGS-THOMSON PERSPECTIVE ON THE MICROELECTRONICS INDUSTRY AND EDUCATION

Strategic Initiatives for Education Systems Outside and Within the Industry for Meeting the Challenges of Our Future

Jean-Claude NATAF, STU and Corporate Training Director
SGS-THOMSON Microelectronics
ZI de ROUSSET
13790 ROUSSET FRANCE

1. Looking Outward

1.1 THE MICROELECTRONICS CONTEXT

The microelectronics industry market today is equivalent in dollars to that of OPEC total oil exports. Yet unlike OPEC, we have not yet reached market maturity and in fact are in the high growth stage of our industry. Our market has grown at over 15% annually over the past 35 years, and this level of growth is expected to continue well into the next century, with predictions for a world semiconductor market of 700B$ in less than 10 years.

That microelectronic components play a vital role in our global economy, from their use in consumer products to semiconductor equipment, is undisputed. Indeed this vast world market will hold many of the future's employment opportunities. We estimate the creation of 500,000 new - and generally high technology - jobs from 1995 to 2000 in the semiconductor and related industries.

This technology driven sector requires a labour force of mostly qualified, highly skilled labour. A major challenge for us and for European education systems will be to bridge the gap between available skills and the jobs of our future.

1.2 WHAT CAN BE DONE TO BRIDGE THE GAP?

A first step, already beginning to happen today, is the direct implication of the microelectronics industry in universities and engineering schools. We must develop strong partnerships, sharing our expertise, resources and feedback.

T.J. Mouthaan and C. Salm (eds.), Microelectronics Education, 7-8.

On a deeper level, the microelectronics industry, and other job creating organisations of the future, must open a dialogue on the needs of our education systems in the coming years. Through alliances with national education systems, governments and communities, we must work together to define the skills and competencies required to succeed in tomorrow's environment and to develop an implementation plan which can respond rapidly to our ever evolving culture. This should deal with educational content but also with methods and culture, and includes skills like the ability to actively seek knowledge, to find innovative solutions, to adapt and to work in collaboration with others.

2 Looking Within

2.1 MANAGING KNOWLEDGE

The educational system is evolving within our own work environment as well. Microelectronics is an industry which is increasingly brain intensive. Knowledge, more than any other factor, is the ultimate resource for our organisation today. Our ability to build new knowledge, capture it and share it so that it flows beyond the individual into the organisation is crucial to remaining competitive in our industry.

2.2 BECOMING A LEARNING ORGANISATION

This necessity is behind the strategic initiative of SGS-THOMSON taken in 1994 to become a learning organisation, and led directly to the creation of our Corporate University, STU. The mission of STU is to provide our people the skills, knowledge and cultural adaptability the company needs to remain abreast of important changes, while strengthening their sense of belonging and entrepreneurial spirit. This learning process is key to our corporate strategy and people focus, and our management is wholly committed to this process.

3 Conclusion

The future of the microelectronics industry will belong to those organisations and people who possess not only core technological competencies and knowledge but who also have the capacity to respond to the needs of a rapidly evolving, complex environment. To achieve success, individuals, companies and even outside institutions, through alliances with industry, must be able to renew or revitalize themselves as part of a process of continuous improvement and adaptation, which can only be achieved through learning.

THE MICROELECTRONICS ENGINEER: EDUCATING FOR THE FUTURE

Marcel J.M. Pelgrom
Philips Research Laboratories WAY52
Prof. Holstlaan 4, 5656AA Eindhoven

1. Changes in industry

During the early years of industrialisation microelectronics companies were essentially units in existing large companies. These companies were structured in a vertical sense: the company would produce its own semi-fabricated articles to fabricate its end products. Also the young microelectronics divisions were structured in this way. However in the mid-80's several start-up companies showed that concentration on particular segment could be more beneficial, this caused a major reordering of the old picture. Now companies no longer try to cover the path from raw materials to consumer goods, the new direction is to specialise on a particular step in that process and dominate the market as one of the few players in that specific field. Figure 1 shows this remarkable change of industry orientation ref. [1].
The European electronics industry is traditionally focussed more on consumer applications than on computing. Due to the globalisation in the consumer world, the above noted process is certainly applicable on the European industry. The previously vertically organised companies have changed in a way where they fit more to the right side of Figure 1, which means that their divisions have become more mutually independent, they try to specialise and become market leader in a segment.

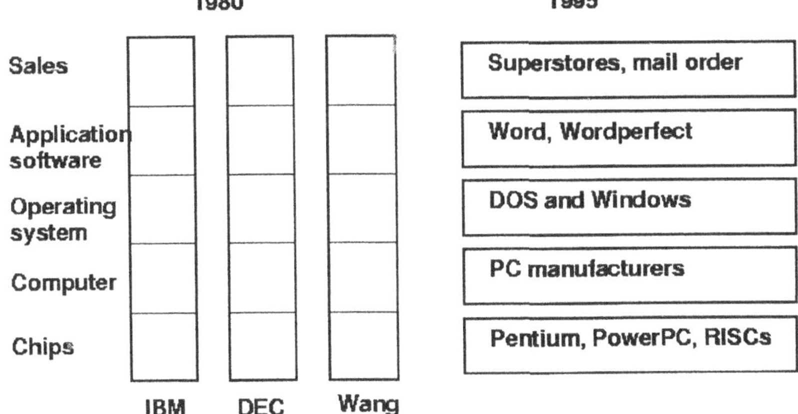

Figure 1 the rotation of vertically oriented industries to specialised organisations [1]

These changes have had many consequences. Firstly the competition between companies shifted from end products to half fabricated goods: ICs are a product on

T.J. Mouthaan and C. Salm (eds.), Microelectronics Education, 9-12.

their own. The same holds for all forms of software and even for Intellectual Property Rights (IPR) a market is opening.
Another major change is the need for short time-to-market. In a situation where only few high-capacity companies fight for the market it is extremely important to be there first. The traditional internal delays in vertical organisations are not acceptable in an open market.
The strong focus on the IC market segment has generated tremendous advances in IC technology. The possibility of millions of transistors on one chip has led to full integration of all system functions on one chip: "systems on silicon".

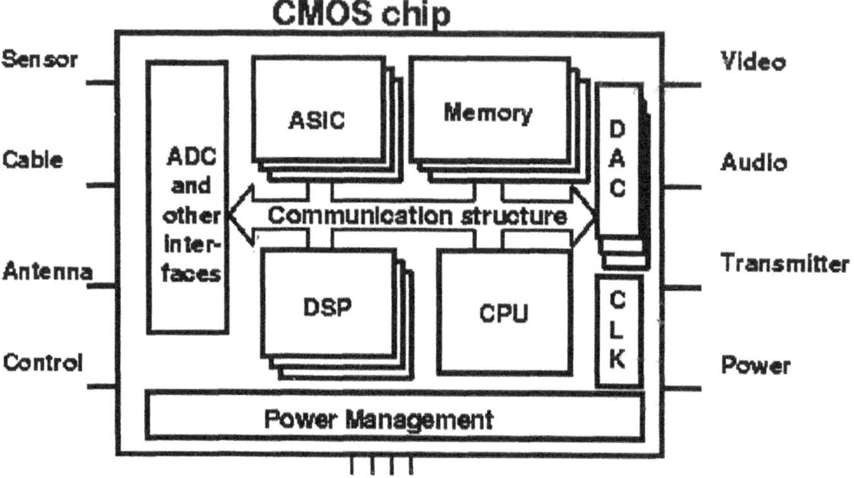

Figure 2 Systems on silicon.

Various techniques are combined to realise all necessary system functions, digital blocks, like memories, processors and special logic are combined with more analogue functions as analogue-to-digital converters, clocking, RF etc. Also a considerable amount of software can be embedded in these chips.
The organisational restructuring and the huge markets for those who can master the technical complexity challenge have changed the organisation of IC design.

2. Organising for systems on silicon

Some ten years ago a designer confronted with a system question could choose one of the available in-house processes, set-up an IC proposal, design the IC and debug it: the designer was the bridge between technology and system. With the complexity of today's CMOS ICs that designer would not finish a single project in his lifetime. Companies have set-up new structures to accommodate the problems associated with handling the IC complexity in such a short time that they can reach for market leadership.
Figure 3 shows the basic organisation form for handling the design of application specific IC products with high complexity that have to be designed in a short time with a high probability of "first time right". Essentially the IC is built up from more

complex blocks or modules that are designed by specialists groups. Examples of these blocks are: memories, analogue interfaces (Analogue-to-digital converters), digital signal processors, and controllers with embedded software, library elements. The design flow is the integrating tool that provides the necessary means for handling all elements.

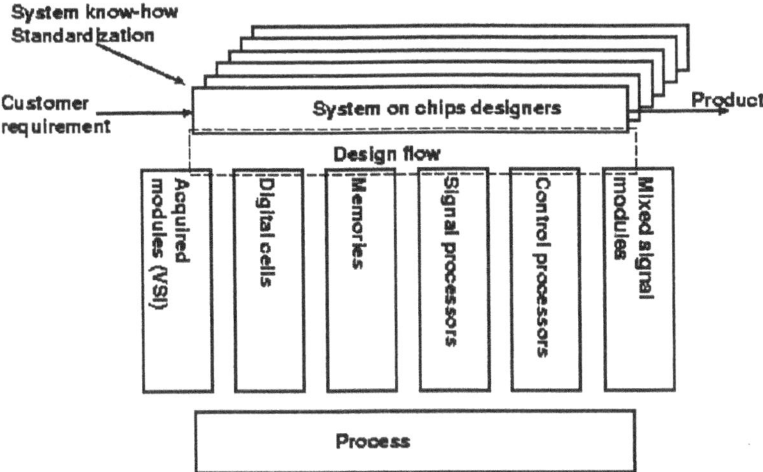

Figure 3 Organisation for the design of high complexity ICs.

IC system architects define the IC functionality based on the customer's wishes. Insight in physical limits of the IC blocks is combined with a system overview. The technical trade-offs between the various blocks and their software content have to be handled, while keeping a clear view on the various market aspects: yield, time-to-market, price, etc. Next to technical aspects also related issues play a role: standardisation, testing, product portfolios etc. It will not be a surprise that industry is diligently looking for designers with system integration capabilities.

Many starting designers are involved in the various module activities: they translate the basic functionality of the CMOS processes into workable units for the system architects. They play a major role in defining the details in a process and must be experienced in handling the design-technology trade-offs.

The single IC designer, who formed the interface between the technologists and the set maker, is rapidly replaced by a number of specialists, each covering a single IC issue. The concentration in specialised layers that is observed on a macro scale in industry is also visible on micro scale in the IC design. The new interfaces create opportunities, but are also a cause of communication problems.

3. Educating for the future.

The microelectronics engineer is a team player that operates as a subcontractor for the multi-man year system-of-chip designs. The skills that are required have not changed fundamentally: basics in physics, mathematics and electronics with elements as (discrete) frequency and time domain, stability, stochastical methods, stability of control loops and a thorough background in experimental skills. Yet, every decade in technology development creates a different set of boundary conditions that put a different emphasis on the various aspects of microelectronics education.

Handling complexity: the 100 bipolar transistor analogue opamp seems incomparable to the 10 million CMOS transistors digital system on silicon. Nevertheless there are similarities in the way designers tackle their problems. The bipolar transistor is a 10-parameter abstraction of a sequence of diffusions, the same idea is followed in very complex chips: larger functional units are abstracted and described by their main parameters. The basic skill involved is to find the dominant functionality and separate it out of the jungle of irrelevant effects. In both cases the experienced designer will oversee a broader range of relevant behaviour and side effects than the novice will. It is not a coincidence that former analogue designers make good system architects.

New systems: systems have become more complex due to the increased possibilities of digital CMOS. A perfect form of storage is available and signals can be manipulated in very non-linear fashions. These basic properties allow introducing a large amount of mathematical techniques into the domain of the system designer. Also today, the field of telecommunication is always a forerunner in introducing new systems. Although these techniques are not too complicated from a mathematical point of view, the trade-off versus the physical boundary conditions is determining their merits. Other important areas that influence the domain of the IC design are innovations in power and storage.

Physics: The trend towards full system integration on chip and the atomic scale of modern transistors require an additional effort in the physics and basic electronics aspects of semiconductor devices, storage media, displays, etc. Quantization effects in nanoelectronic structures are like to become a major bottleneck.

Software: The algorithmic and software sides of ICs are of great importance. This disciplines allows to tackle complexity of the large number of elements in an IC and generate the abstractions required to implement digital functionality. However, software is always related to the physical world, therefore in education software should never be a subject on its own.

Team workers: the high interaction between various disciplines in a team that realises complex chips puts additional demands on the social skills of its members. Efficient communication, in writing, in presentation, in informal contacts and some feeling for multicultural teams, is often a decisive factor for the success of a team.

4. Conclusion

Microelectronics industry is in a process of rapid changes. Technical and commercial boundary conditions require a different attitude of the microelectronics engineer towards the business. A few potentially important topics in microelectronics education have been discussed.

5. References

[1] A.S. Grove, *Only the paranoid survive; how to exploit the crisis points that challenge every company and career*, A Currency book published by Doubleday, 1996.

A REPORT FROM A JAPANESE MICROELECTRONICS COMPANY CONCERNING THE EDUCATION OF INFORMATION TECHNOLOGY

YUMINOSUKE YANO
Fujitsu Computer Technology Limited
2-15-16 Shinyokohama, Kohoku-ku, Yokohama 222-0033, Japan

1 Abstract

The education of information technology so called "Informatics (Informatik)" has been based on various types of process to set up in universities and institutes. The curriculum depends on various types with the kind of establishment. There are three styles consisting of the hardware, the software and both.
This paper describes a report from a Japanese design and development company for the microelectronics technology concerning the education of information technology. The recent status for the establishment of the faculty and department, the relation between the information technology department and others, the typical curriculum and some activities for the development of LSI on microelectronics education in universities are described in order.

2 Introduction

2.1 NEEDS FROM SOCIETY

The industry and company needed excellent information engineers for their EDP application programs. Many software houses were established in 1980's and they also needed many information engineers, especially for computer programmers. Many industries and companies have aimed to grow up to be on the information technology.
The government suggested a lack of many information engineers related with computer programmers in 1980's.

2.2 NEEDS FROM STUDENT

Many students have been interested in the information technology, especially for computers and communications, and they intended to study the information technology in universities and institutes. They also intended to enter into business related with the information technology.

T.J. Mouthaan and C. Salm (eds.), Microelectronics Education, 13-16.

3 Recent Status

3.1 NAME OF FACULTY

Normally the department related with the information technology is included almost in the faculty of engineering and sometimes in the faculty of science or science and engineering. However, some universities and institutes establish special faculties like "Faculty of Information" of Shizuoka University, "Faculty of Computer Science and Systems Engineering" of Kyushu Institute of Technology or "Faculty of Engineering Science" of Osaka University. Some universities use the school or college instead of the faculty. The special name is sometimes used to identify the character of the faculty like "Faculty of Electro-Communications" of the University of Electro-Communications, "Faculty of Engineering and Design" of Kyoto Institute of Technology (including "Faculty of Textile Science"), "Faculty of Computer Science and Engineering" of the University of Aizu (public), "Faculty of Environmental Information" of Keio University (private) or "School of High-Technology for Human Welfare" of Tokai University (private).

3.2 NAME OF DEPARTMENT

There are various kinds of name for the department related with information technology. The popular name of the department is the information engineering in the faculty of engineering or science and engineering. The information science is normally used in the faculty of science or science and engineering. The computer science is sometimes used instead of the information engineering in the Faculty of Engineering of Iwate University, Tokyo Institute of Technology and Chiba Institute of Technology (private). Usually, the Japanese word "Johokogaku" is often used for the information engineering group and "Johokagaku" is used for the information science. Some universities and institutes use to identify the character of the department like "Computer Science and Systems Engineering" of Muroran Institute of Technology, "Information and Computer Sciences" of Saitama University, Chiba University, Osaka University, Osaka City University (public) and Kagoshima University, "Computer Science and Systems Engineering" of Miyazaki University, "Computer Engineering" of Nihon University (private), "Information Science and Engineering" of Kanagawa Institute of Technology (private) or "Information and Computer Engineering" of Kanazawa Institute of Technology (private).
Some national universities combine the information engineering with knowledge or intelligent systems like "Knowledge-based Information Engineering" of Toyohashi University of Technology, "Information and Knowledge Engineering" of Tottori University, "Information Science and Intelligent Systems" of the University of Tokushima and "Computer Science and Intelligent Systems" of Oita University.

3.3 RELATION BETWEEN INFORMATION TECHNOLOGY AND OTHERS

The relation between the department of information engineering or science and others is shown in Figure 1.

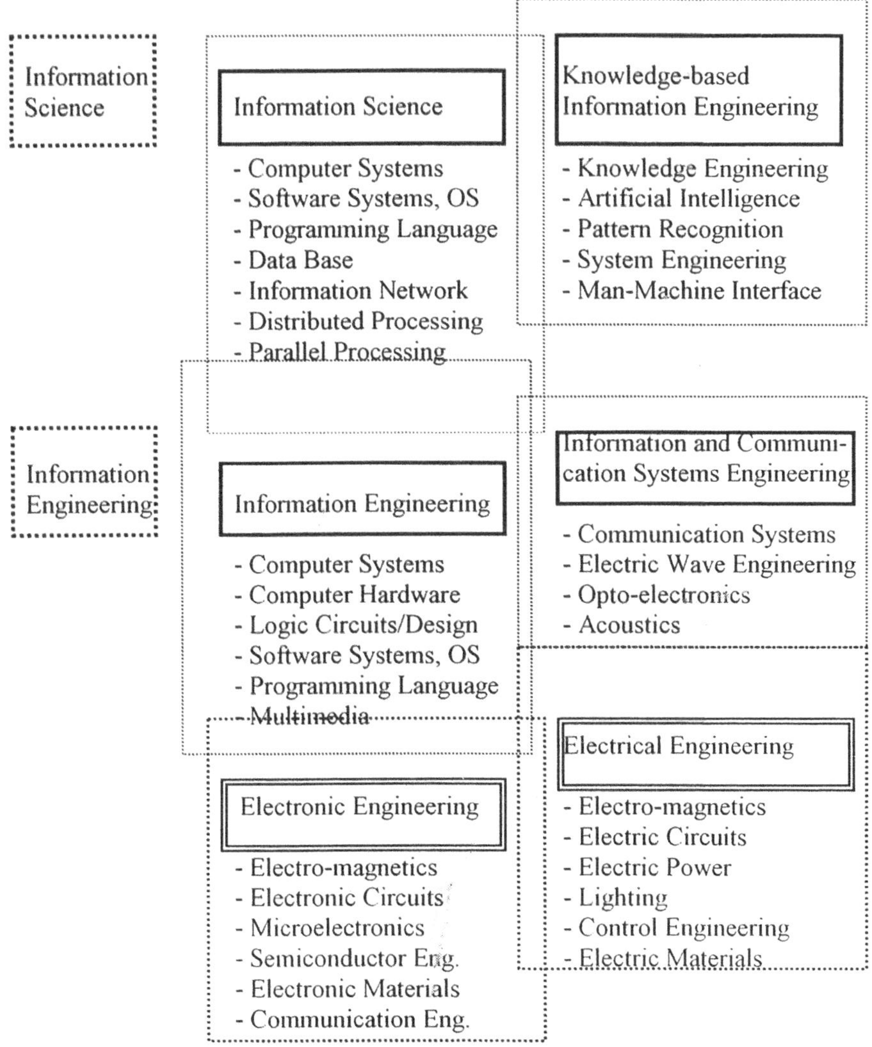

Figure 1 Relation between the department of information engineering or science and others

4 Curriculum

4.1 MODEL CURRICULUM FOR INFORMATION ENGINEERING

"Proposals and Curriculum Recommendations for B.A. Degree Programs in Computer Science and Engineering" has been made by CS Working Group of the Curriculum Committee for B.A. Degree in Computer Science and Information Engineering, Information Processing Society of Japan, in 1990 [1]. This "CS Curriculum J90" is shown in Table 1. The curriculum refers to ACM curriculums 68, 78 and 88.

Table 1 CS Curriculum JS90

Core Curriculum		Advanced Curriculum	
JCS1	Programming: Introduction	JCS8	Operating Systems and Architectures II
JCS2	Programming: Design and Implementation	JCS9	Files and Data Base Systems
JCS3	Computer Systems: Introduction	JCS10	Artificial Intelligence
JCS4	Basic Computer Hardware	JCS11	Human Interface
JCS5	Information Structures and Algorithms	JCS12	Modeling and Algorithms of Computers
JCS6	Operating Systems and Architectures I	JCS13	Software: Design and Development
JCS7	Structure of Programming Languages	JCS14	Programming Languages: Theory and Practice
		JCS15	Numerical Calculations: Theory and Practice

4.2 ACTIVITIES FOR LSI DEVELOPMENT

The design, evaluation, and manufacturing if applicable, are very important curriculum for the microelectronics education. Some universities have developed specific LSI using the service of design education centers, CMP or the individual fabrication center. Three recent activities are introduced as follows.

4.2.1 *Asynchronous microprocessor*

Professor Nanya's group of Tokyo Institute of Technology has developed the asynchronous microprocessor "TAITAC-2" using the service of the VLSI Design and Education Center established at the Tokyo University in 1996.. This microprocessor uses asynchronous logic circuits consisting of 500k transistors with MIPS R2000 architecture.

4.2.2 *Collision detection VLSI processor*

Professor Kameyama's group of Tohoku University has developed the collision detection VLSI processor using the service of CMP in 1996. This VLSI processor consists of 250k transistors with specific bit serial pipe-line architecture.

4.2.3 *Electron Device Research Center*

Toyohashi University of Technology has established the Electron Device Research Center in 1993. The center has fabrication facilities like the ion implanter, clean draft, reactive ion etching systems, electron beam lithography system, mask aligner, diffusion furnaces and EB LSI tester, and the CAD and simulation systems are also installed.

5 References

[1] S. Noguchi et al. "Journal of Information Processing Society of Japan, Vol.32, No.10, pp.1079-1092, 1991

Session P1

INDUSTRIAL PROJECTS/
NEW CONCEPTS

PROFESSIONAL DEVELOPMENT MICROELECTRONICS TRAINING IN EUROPE

Professor Nihal SINNADURAI
Principal Consultant, TWI
Abington Hall, Cambridge CB1 6AL, UK

1. Introduction

The microelectronics markets around the world have undergone major shifts as some technologies have receded and others have gained ground[1]. Consequently, engineers and managers in industry must continually update their professional skills by means of continuing education, to keep up with the fast changing markets and opportunities.

2. Microelectronics Markets[1]

Examples of microelectronics growth include the overall upsurge in semiconductor markets in Europe (Table 1), through concerted actions by manufacturers and the European Commission's (EC) sponsored research programmes, despite some short-term falls in 1997.

Table 1. European Semiconductor Markets (Million ECU)

1994	1998
19800	45700

The markets for hybrid microelectronics grew at an explosive rate during the period 1980-1986. The situation began to change dramatically from 1986, as hybrid functions were increasingly replaced by custom MSI and VLSI circuits. The overall market growth rate for hybrid microelectronics consequently diminished over the past decade (Table 2). The technology analysis shows a continuing erosion of the thick film markets as Surface Mount Technology (SMT) and thin film markets have gained ground.

Table 2. Total European Market for Hybrid Circuits (Millions ECU)

1986	cagr	1990	cagr	1994	cagr	1998
1178	15%	2066	9%	2900	8.0%	3950

The thick film segment of the market (Table 3) shows a decline in growth in contrast with a significant growth in thin film circuits. The thin film segment has since maintained steady growth driven by the increasing requirement for interconnection and integrated thin film drive circuits for LCDs and MultiChip Modules (MCMs). Meanwhile, Surface Mounting Technology (SMT) whose growth during 1986-1990 was 100% cagr (compound annual growth rate) and is now about 12%, has become the dominant assembly technology for PCBs and the alternative technology for hybrid

T.J. Mouthaan and C. Salm (eds.), Microelectronics Education, 19-22.

microcircuits. It is clear that SMT will be the dominant technology for the remainder of this decade and well into the next.

Table 3. Technology Distribution of European Hybrid Microelectronics Market (Millions ECU)

TECHNOLOGY	1986	cagr	1990	cagr	1994	cagr	1998
SMT on PCB ≤6"x4"	35	100%	558	21.5%	1216	11.8%	1900
Thick Film	1084	4%	1364	1%	1419	1.4%	1500
Thin Film	59	16%	145	16%	265	20%	550
TOTAL	1178		2066		2900		3950

The shift to, and current dominance by, SMT (Figure 1) is reinforced by the massive shift to surface-mount packaging of integrated circuits.

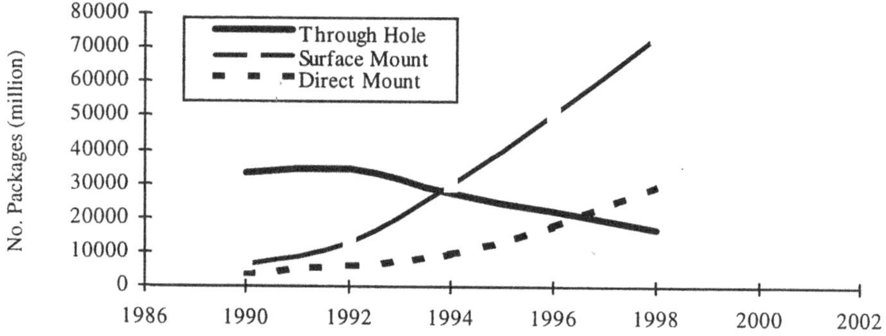

Figure 1. Semiconductor Device Package Market Trends

The re-emergence of MCMs as a solution for compact electronics technologies creates new training needs (Table 4). MCM potential is in both terrestrial and space applications.

Table 4. MCM Market Growth, Europe, 1990-1998 (Value)

Year	1990	CAGR	1994	CAGR	1998
Million s ECU	260	21%	550	18%	1100

3. Professional Development Training Provision in Microelectronics

Awareness of the actual industrial markets and future changes are essential intelligence both for industry and for those who wish to serve the training needs of industry. Indeed, the communication of such intelligence to undergraduate and graduate students is necessary to help and encourage the students to make informed decisions about educational choices that will prepare them for careers in industry. Such an informed, market-driven approach does not yet exist. Both academic and industrial training are provided either according to the existing engineering skills of those delivering the training or in response to the preconceptions of undergraduates enrolling for courses. The subsequent survival of the courses is market dependent. The courses are more likely to survive if they are constructed using market intelligence.

Pan-European programmes for professional development training have tended to be conceived from a perception of a need for training followed by significant focused action to deliver to that perceived need. Such programmes include EuroPACE sponsored by major corporations to provide training for in-house needs by means of video training courses transmitted from Paris via EutelSat to company sites in Europe. After many hundreds of transmissions over seven years, EuroPACE was ceased because the training was not adequately used. The concept was good, and leading experts provided the lectures. However, despite the apparent flexibility of using recorded material for distance learning, the majority of users did not adapt to this method of learning. A lesson learned is that training should be suited to the users rather than requiring the users to adapt to the method of teaching.

Another training initiative was EuroPIC, which intended to adapt in-house company training from one organisation to provide training materiel for use by other company trainers. In this programme, existing training materiel on all aspects of semiconductor production, was to be translated from Dutch into English and supplemented by practical training aids. The objective was to train company training departments and provide the materiel for in-house training. Despite the translation of a large part of the original materiel, and active promotion to major companies, there was insufficient interest by such companies, and the programme was terminated. The lesson learned is that courses tailored to one company and not otherwise tested in the open market, are not necessarily suited to other companies' needs. EuroChip was another EC partnership initiative to provide academic institutions with low-cost access to semiconductor design, fabrication and training, in order that European competitiveness would be increased by instilling such skills in the universities and students. The programme was successful within academia and enabled many students to learn useful design skills and apply them in their research projects. The latest thinking is that technology access and training should be facilitated directly for industry, and especially for small and medium sized enterprises (SMEs), in a manner which responds to their needs in the marketplace.

EuroPRACTICE is the latest EC partnership initiative intended to provide European industry with direct access to design, technology, fabrication, testing and training. Differently from previous initiatives, it aims to facilitate access to services from commercial and academic suppliers of ASICs, MCMs and Microsystems (µS). To meet this objective, the EuroPRACTICE Training and Best Practice Service (TBPS) set out to locate good Continuing Professional Development (CPD) providers in Europe and to match the CPD training to the needs of industry. An early task was to study the market needs for training[2]. An easy questionnaire was distributed to 38,000 electronics engineers and managers throughout Europe, and over 1000 replies were received in the proportions shown in Figure 2 and analysed as shown in Figure 3.

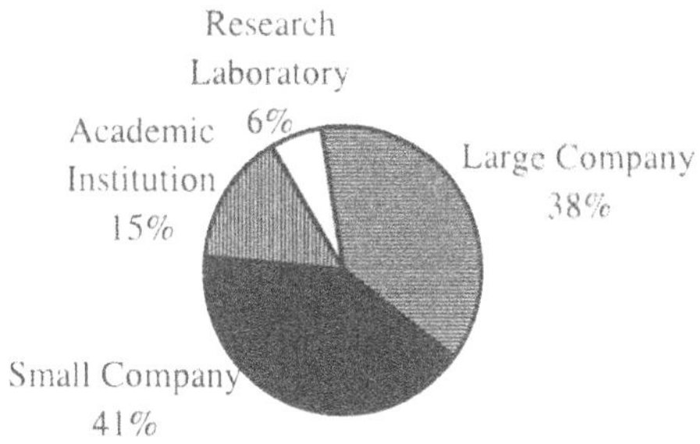

Figure 2. Organisations Responding to Training Needs Survey

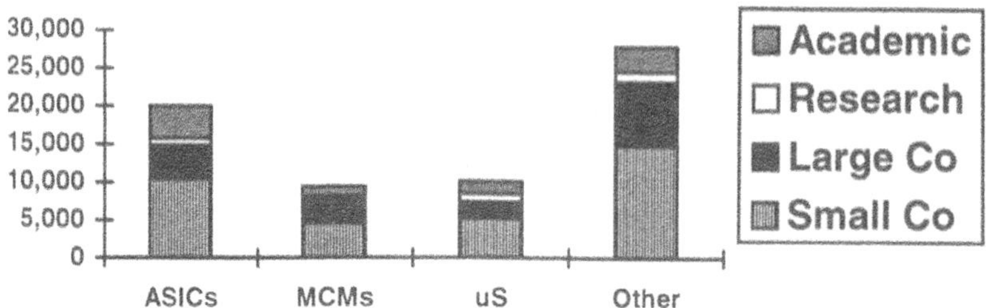

Figure 3. Estimated European Need forTraining

The survey received responses from about 15% of relevant companies in Europe. The analysis indicates that up to 10,000 organisations would benefit from CPD training. The dominant demand fell into the "Other" category, comprising training in packaging, reliability, assembly, project management and related subjects. Such an outcome is not surprising, considering the technology market research described earlier. The need for ASIC training is equal to that for MCM and μS combined. While there appear to be sufficient ASIC courses, there is a large shortfall in available MCM courses and a lesser shortfall in μS courses. Another market analysis[3] shows a significant growth of 6%-15% p.a. in the market for mixed signal applications, and also that there is a serious shortage of analogue designers. Companies are arranging their own application oriented training to create their own base of skilled people, because academic courses are too remote from the practical needs of industry, and there are too few graduates taking up the available university courses. There is a distinct opportunity to provide industry-focused CPD training in analogue design.

In Europe there are more than one thousand short courses in microelectronics provided annually by commercial professional development organisations and universities. Courses in microelectronics have been provided for some twenty years, and have survived and grown by adapting to market needs. The challenge remains to reach the people needing the training and to facilitate their attendance at the training courses. Studies[4] of SMEs investment in future development indicate that 40% will be out of business in 5 years, 50% are static and only 10% are likely to grow, with only 4% within that as real investors in forward development. The challenge for training providers is to identify the 4% of SMEs who would respond, and to adapt the type and duration of the courses to their needs. Courses range from one day to a few weeks and include publicly advertised scheduled offerings and custom in-house courses requested by companies. Analyses show there is a preference for two-day courses within easy reach of the companies. Both hands-on (best practice) and lecture courses are needed. Also desired is European CPD accreditation, so that professional engineers can accumulate professional development units (PDUs) towards Euro Ingenieur (Eur. Ing).

4. References

1. "Microelectronics Markets in Europe, Supplemented by World Data", ATAC, 1997
2. T Cartwright "Market Survey of Potential Users of Training" EuroPRACTICE Training and Best Practice Service, TWI Report T/1296/023, February 1997, as reported in EuroTraining News, Issue 3, December 1996.
3. N R Stockham, C J Condie, N Sinnadurai, "Analogue Design Training Needs in Europe - a Preliminary Study", EuroPRACTICE Training and Best Practice Service, Report, June 1996, , as reported in EuroTraining News, Issue 3, December 1996.
4. D. Storey, "Ten Percenters - Fast Growing SMEs in Great Britain", 1997.

INITIATIVE TO ADDRESS THE SKILLS SHORTAGE IN THE MICROELECTRONICS INDUSTRY IN IRELAND

B. O'NEILL, J. DONNELLY, C. KELLEHER, G.T. WRIXON
NMRC, Cork, Ireland.
J. LINEHAN, P. O'CALLAGHAN
FÁS, Cork, Ireland.
O. DUGGAN
NCVA, Dublin, Ireland.

1 Abstract

The electronics industry, and the Microelectronics industry in particular, has experienced massive expansion in Ireland over the past 10 years. As well as the economic benefits that this expansion brings it also creates the need for a supply of well trained people at all levels.

In Ireland there is a shortage of people with specific semiconductor training for the Microelectronics industry. This shortage is often exacerbated by multinational companies over specifying the educational requirements for certain jobs. Ireland has, by European standards a young and highly educated population. In a novel initiative between the National Microelectronics Research Centre (NMRC) and the National Training Authority (FÁS) a programme was set up to take people with qualifications at or below technician level and train them specifically for work as operators or technicians in the microelectronics industry.

This paper describes the industrial climate in Ireland which brought about the need for this programme and how a focused initiative involving co-operation between state funded organisations can be used to solve some of the problems facing the microelectronics industry.

2 Introduction

Ireland has one of the most dynamic economies within the EU, particularly in the area of attracting inward investment from the USA. One third of all electronic and healthcare greenfield investment into Europe [1] from the USA goes to Ireland. More than half of all new software investment into Europe goes to Ireland and 23% of all US manufacturing industry investment in Europe goes to Ireland [2]. This makes Ireland the fastest growing European economy in the area of electronics, it is the most frequently selected location for leading edge technology and for low cost as shown in figure 1 [3]. It also provides investors with the highest return on their investment.

T.J. Mouthaan and C. Salm (eds.), Microelectronics Education, 23-26.

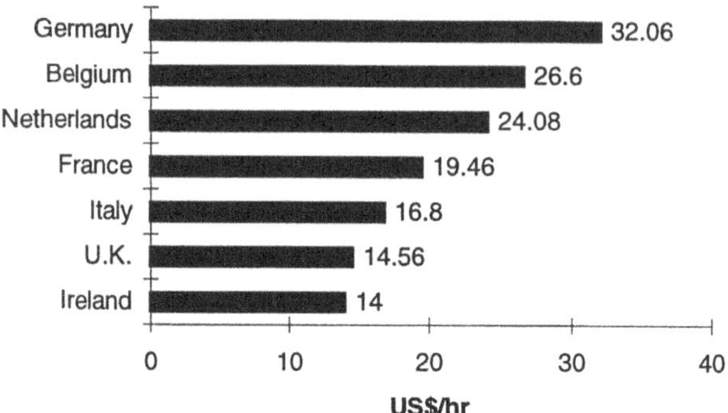

Figure 1 International comparison of payroll costs

3 Youth and Education

Contributing to this success on the economic front are the Irish demographics and the Irish educational system. Ireland has one of the youngest populations of the EU, 44% of the Irish population are under 25, the indexed supply of young people for Europe is shown in figure 2 [4]. The educational system is highly regarded by incoming investors and 63% of the population participate in education to at least the age of 18. The emphasis in the third level educational establishments is towards technology and business with 60% of third level students involved in Engineering, Science or Business Studies.

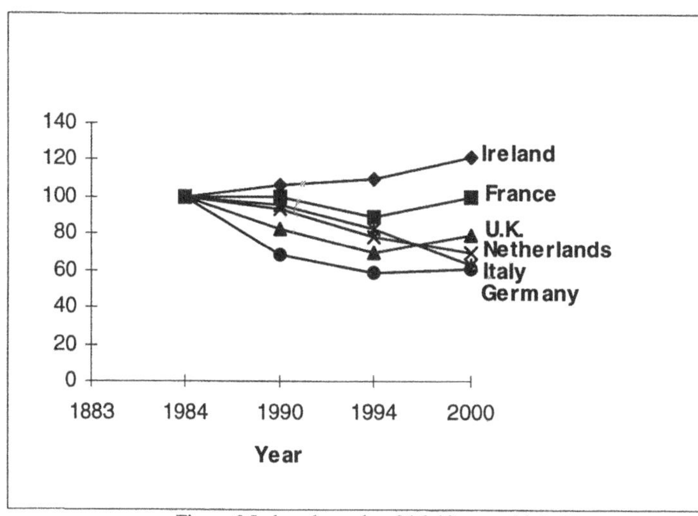

Figure 2 Indexed supply of 15-19 year olds in Europe

4 Government Policy

The Irish Government has a commitment to maintain this growth through programmes designed to meet the skills needs of Ireland's Overseas companies. In March of 1997 the Government announced a strategic approach to meeting these needs. The main points of this strategy are:

1. To increase the output of computer software graduates with an additional 1200 third level places available in 1997/1998

2. An additional 1000 multi-lingual educational places available with temporary placement abroad for students.

3. The establishment of fora involving the National Training Authority (FÁS) and local third level colleges to design innovative means of meeting local needs particularly for the electronics industry.

4. To make a further 750 places available for electronics technicians in the Regional Institutes of Technology.

5 Semiconductor Training Programme

This programme came about as a direct result of the government initiatives described above. The objective of the programme is to provide a pool of people trained in skills directly related to those necessary for operative and technician positions in the microelectronics industry in addition to their more formal qualifications. This then would fulfil two needs, it would open up the industry to a wider range of people and provide for the industry a pool of potential employees with a skill set which matches their requirements.

In meetings between the Southern Regional Director of FAS and the Director of the National Microelectronics Research Centre it was decided to run a pilot programme to train people from the register of unemployed specifically for work in the electronics/microelectronics industries.

The programme has four separate phases:

Basic Education Module

To bring those who might have been out of formal education for some time back up to a basic level of understanding in mathematics, physics, chemistry and basic electronics

Semiconductor Module

This module was carried out at the NMRC facility and covered all of the topics related to semiconductor fabrication, cleanroom technology, facilities, oxidation, diffusion, implantation, etching, CVD, metallisation, and process integration. About 50% of the time on this four week module was spent in the cleanroom working on process and measurement equipment.

World Class Manufacturing Module

Most successful companies in the semiconductor sector operate World Class Manufacturing systems. Topics related to this area, such as Total Quality Management(TQM), Total Productive Maintenance(TPM), problem solving, teamwork skills, ISO9000 and ISO14000, were taught by FÁS in this module.

Work Experience

The trainees were sent out to companies operating in the electronics sector for 4 weeks work experience.

The trainees selected for the initial course were well motivated and extremely interested in making a career in the microelectronics industry. The NMRC has a background in providing custom tailored training courses for industry at a variety of levels from general operatives right through to post graduate level. Much of the material presented to these trainees has in the past been presented to higher levels. Such was the motivation and interest from these trainees that the interaction between the trainers and the trainees during both the classroom and practical sessions was on a par and better than on many of the courses to better qualified attendees.
The course has also been recognised by the National Council for Vocational Awards (NCVA) as an elective module towards the National Vocational Certificate which is a European level 2 qualification.

6 Summary

The shortage of trained people for the microelectronics industry has been addressed in two ways, a commitment to third level education and the provision of a path for unemployed people to achieve the skills necessary to make them attractive as employees to the industry. This paper has given the background of Irelands recent economic success in the area of high technology industry and has shown how a national policy has been put in place to address the needs of these industries. One of the programmes under this policy has been described in detail, the success of which can be gauged by the fact all of the trainees from the pilot programme have been employed in the electronics industry

7 References

[1] E. Schumann, Channel, Volume 11, number 2 p. 14.
[2] *E&Y International/Buck Consultants
[3] *Swedish Employers Confederation, 1996
[4] *IMS/Eurostat
* Private communication from Irish Development Authority (IDA)

DESIGN OF A LOW VOLTAGE/LOW POWER MIXED ANALOG/DIGITAL ASIC IN CONTINUING EDUCATION

L. HEBRARD, H. EL KHANTOUR, C. PETER and F. BRAUN
ULP - Pôle CNFM MIGREST - LEPSI
23, rue du Loess B.P. 20 - 67037 STRASBOURG - Cedex 2 (FRANCE)
http ://www-lepsi.in2p3.fr

1 Abstract

From our recent experience on training in ASIC (Application Specific Integrated Circuit) design, this paper shows how SMEs (Small and Medium sized Enterprises) can benefit from continuing education to enter the field of microelectronics in order to add ASICs in their products. A low voltage/low power mixed analog/digital ASIC was designed by a student during the one year final project required to achieve the engineering degree prepared through the French CNAM (Centre National des Arts et Métiers) continuing education program. This circuit is intended to domotic applications and we show how such a training has lead the SME to continue towards an industrial version of this ASIC. The introduction of this new ASIC into the current products of the SME is foreseen for october 1998.

2 Introduction

SMEs not involved in microelectronics see the introduction of ASICs in their product often as a highly risked forward step, especially when the ASIC has to be a low voltage/low power mixed analog/digital ASIC. In this former case, FPGAs (Field Programmable Gate Arrays) can't be used and a full custom solution is the only alternative. Two main reasons explain the feverishness of SMEs to go ahead. The first one comes from the investment required. Although the expected return on investment is often high (better reliability, less components than a Printed-Circuit-Board solution, lower cost for the final product assembly,...), the first investment for prototyping is equally high in comparison to classical solutions (PCB). So the first prototype has to be the good one! The second reason comes from the lack of ability in the field of microelectronics. Even when the SME doesn't think to design by itself the ASIC, it must have engineers able to understand microelectronics in order to discuss efficiently the specifications of the ASIC with the subcontractor involved in the design of the circuit. This is a key point to ensure the successfull working of the first prototype. At this level,

T.J. Mouthaan and C. Salm (eds.), Microelectronics Education, 27-30.

continuing education is the best way for SMEs to educate their engineers for managing an ASIC design project.

3 ASIC description

The ASIC designed for this project is dedicated to a domotic application. Confidentiality reasons prevent us from giving a detailed description of the circuit. Nevertheless, the block diagram of figure 1 shows that the ASIC is split into three main parts.

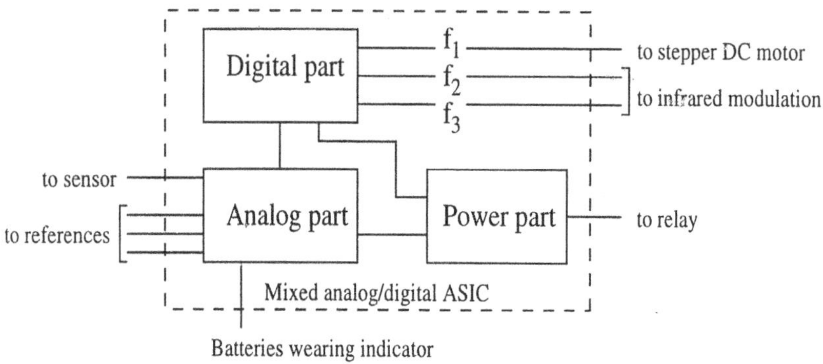

Figure 1 : ASIC block diagram

The power part is intended to drive an external bistable relay. The analog part compares references to external domotic measurements and provides the order to switch the relay. It controls also the power consumption through a band-gap voltage regulator [1] which limits the power supply voltage to 2.2V. Finally, it detects the battery wearing in the case of a supply through batteries (2x1,5V). The digital part controls an analog input multiplexer and provides clock signals at different frequencies. These signals modulate an infrared signal used to switch a remote-operated actuator. They are also used to drive the stepper DC motor of an external clock.

4 First results

The circuit was designed using the AMS1.2µm CMOS technology and has been functionally characterized with success. The mean power consumption including the power part (stepper DC motor and bistable relay) is well under 200µA which allows for a high batteries life time. Figure 2 shows the circuit layout. It is 1800µm width and 2500µm long. On the upper rigth corner, we recognize the digital part and on the left, the analog part. The power part stays at the bottom.

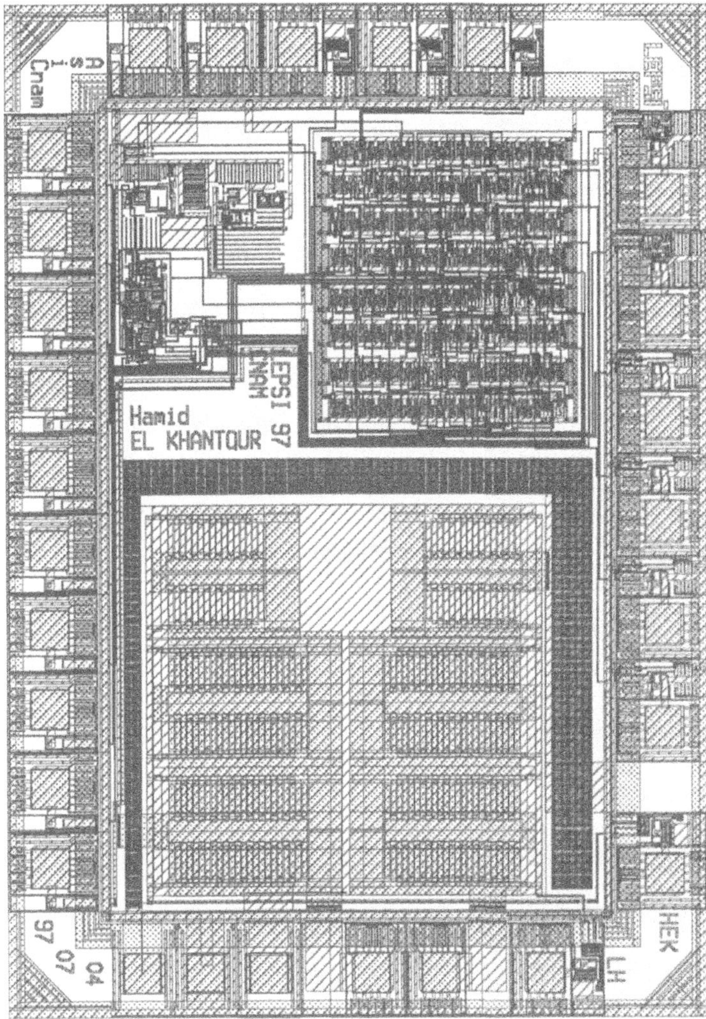

Figure 2 ASIC layout

5 From continuing education to ASIC industrialization

The circuit was designed during the one year final project required to achieve the engineering degree prepared through the French CNAM continuing education program. After the successfull results obtained [2], the SME involved in this project decided to study the industrialization of the ASIC. Two alternatives were considered. The first one, which was to kit itself out with microelectronics CAD tools and to design by itself the industrial prototype, was quickly dismissed. SMEs have to design several ASICs a year to write off the cost of CAD tools and design kits. This is merely the case! So the second and chosen alternative was to subcontract the ASIC design. Even if abilities in microelectronics are not absolutely necessary to become involved in such a project, the knowledge acquired by the SME through this continuing education program in microelectronics was very important to discuss efficiently the specifications of the ASIC to be designed. It was also a key point for the SME to decide to introduce the ASIC in its products, and will contribute without doubt to the success of the first industrial prototype foreseen for the end of the first semester of 1998.

6 Conclusion

The introduction of mixed analog/digital ASICs in conventional products is seen as highly risked by most of the SMEs not involved currently in microelectronics. To be confident in the benefits they can obtain from such a solution, SMEs have to acquire abilities in the field of microelectronics. These abilities can lead the SME to design by itself the ASICs it needs, or more often to subcontract efficiently the design. We have described our recent experience in this field, and we hope it will lead other SMEs to follow the same way.

7 References

1. K. E. Kuijk, « A Precision Reference Voltage Source », IEEE JSSC, pp. 222-226, June 1973
2. H. El Kantour, « Mémoire d'ingénieur CNAM », to be published on 1998

8 Bibliography

3. P. Allen, D. Holberg, ``CMOS Analog Circuit Design", *Ronehart and Winston*, 1987
4. K. Laker, W. Sansen, ``Design of Analog ICs and Systems", *McGraw-Hill*, 1994
5. N. Weste, K. Eshraghian, ``Principles of CMOS VLSI Design", *Addison-Wesley Publishing Company*, 1993

THE LIMITS OF THE ELECTRICAL NEUTRALITY HYPOTHESIS

A PEDAGOGICAL APPROACH

Georges KAMARINOS
LPCS-ENSERG
23, rue des Martyrs, BP 257
38016 GRENOBLE Cedex 1, France

1 - Introduction

The classical (usual) teaching of Electron Devices Physics uses constantly the hypothesis of the Electrical Neutrality (E.N). This assumption gives always the most basic equations for the analysis, the modeling and the simulation of the working of the Electron Devices and the Integrated Circuits.

Indeed :

(i) the E.N gives the equation of the dependence of the Fermi level on the temperature in the substrate material.
(ii) The E.N of the global Space Charge Region in the Schottky model, leads easily to the ideal current -voltage characteristics of the P-N junction;
(iii) the E.N concerning all parts in a MIS structure gives one basic equation for every such device (MIS Capacitance, MIS Tunnel diodes, different Memory Structures, MOS Transistors...).

In spite of the productiveness of this hypothesis, it is useful to remind that it is only a (good) approximation and, now, the situations where it fails become frequent.

In this paper the author points out the principal causes of the failure of the E.N hypothesis and he presents them in an attractive and simple pedagogical manner.

2 - Electrical Neutrality and Injection

The injection of carriers is the basic effect for the working of bipolar devices. The charge relaxation, deduced from Maxwell equations, in a homogeneous material, of resistivity ρ and dielectric constant ε, is given by:

$$\tau_d = \varepsilon\rho \qquad (1)$$

For usual semiconductors, serving to microelectronics technology, τ_d is very short: on the order of 1 ps for 1 Ω cm Si at T =300 K. Evidently this is not the case for «new» materials, like diamond of SiC, or even insulator substrates in SOI and TFT's techniques.
It is so important to point out that for such materials and, also, for very fast devices the Space Charge Limited Current regime SCLC [1] p. 102, is present and it can become dominant.

T.J. Mouthaan and C. Salm (eds.), Microelectronics Education, 31-34.

In the majority of situations, the so called double injection regime is valid ; the profile of density, in time and space, of the neutral electron-hole plasma is often given by solving the simple Ambipolar Diffusion Equation (A.D.E) ([2] p. 83 ; [3] § 8.7)

$$\delta n \cong \delta p; \qquad -\frac{\delta p}{\tau} + \bar{E}\,\frac{\mu_n \mu_p (p-n)}{n\mu_n + p\mu_p}\nabla p \;+\; \frac{D_n D_p (p+n)}{D_n n + D_p p}\nabla^2_p = \frac{\partial p}{\partial t} \tag{2}$$

The A.D.E is a linear combination of the hydrodynamic continuity equations for electron and holes; in these equations, the fluxes of each type of free carriers are monokinetic:

$$\bar{V}_n = -\mu_n \bar{E} \;;\; \bar{V}_p = \mu_p \bar{E} \tag{3}$$

Under conditions of strict E.N ($div\,\bar{E} = 0$), the two continuity equations are compatible <u>for a unique value</u> of the electric field given by ([4] p. 11):

$$E_0^2 = \frac{(\mu_n - \mu_p)^2}{\mu_n + \mu_p} \cdot \frac{1}{2\tau\mu_n\mu_p} \cdot \frac{kT}{q} \tag{4}$$

and thus the profile of free carriers density, in one dimension, is given by:

$$\delta p = \delta n = (\delta p)_0 \exp - \frac{x}{\tau\mu' E_0} \;;\; \mu' = \frac{2\mu_n\mu_p}{\mu_n - \mu_p} \tag{5}$$

Evidently, the free carriers have velocities which are distributed according to Boltzmann statistics and the strict E.N. never holds, even for E_o.

Besides the SHR dominant G - R kinetics require two different life times τ_p and τ_n and consequently $\delta n = \delta p + \delta n_r$ (δn_r are the charged traps) and the A.D.E (2) does not hold.

The E.N. is then only an approximation which has to be tested every time when A.D.E is used (in bipolar devices) and when a detailed profile of fields and densities is required.

3 - The Space Charge Region Length

3.1 THE INTEGRATION OF POISSON EQUATION

The extension of the Space Charge Region (SCR) in a device is a basic parameter for its modelization and the simulation of its working. The solution of Poisson equation:

$$\varepsilon \, div\, \bar{E} = \rho_v \tag{6}$$

gives the length of SCR..

The so called Maxwell-Boltzmann solution gives the first integral of equation (6) [5,6]

$$E(x) = \frac{kT}{q} \frac{\sqrt{2}}{L_D} \sqrt{\exp[U(x)] - U(x) - 1} \tag{7}$$

Where L_D is the extinsic Debye length

$$L_D^2 = \frac{\varepsilon kT}{q^2 N_D} \tag{8}$$

The second integration of (6) is possible only numerically.

Nevertheless simple approximative solutions exist for the two extreme cases of (i) very small and (ii) very large E.N perturbations [7]:

$$\text{(i)}\quad U(x) = U_s \exp\left(-\frac{x}{L_D}\right) \tag{9}$$

for very small perturbation,

$$\text{(ii)}\quad U(x) = 2\ln\left[\exp\left(\frac{1}{2}U_0\right) + \frac{x}{L_{sc}}\right] \tag{10}$$

for very large perturbation.

L_{sc} is a screening length equal to $\sqrt{2}\left(1 - \sqrt{N_D / n_s}\right) L_D$.

U_0 is the reduced surface potential and n_s the surface density of electrons.
The above potential fonctions are used in the great majority of modeling of usual microelectronic devices (p-n junctions, MIS structures, MOS transistors [8]).

3.2 THE INFLUENCE OF FREE CARRIERS AND DEEP ELCTRONIC LEVELS

(i) It is easy to show that the solution of the Poisson equation, in a semi-infinite homogeneous semiconductor, taking into account only the shallow dopant density N_D, gives a well defined SCR length:

$$W = \sqrt{\frac{2\varepsilon_S \Phi_S}{q N_D}} \tag{11}$$

ϕ_s is the surface potential..

(ii) When the density the majority of free carriers is taken into account, $\rho_v = q\,(N_D - n)$, an exact definition of W is not possible; indeed theoretically the SCR length becomes infinite. A possible, empirical, definition of W can be given as the length x_0 where the ratio of remaining space charge Q_r to the total space charge Q_t is lower than a value a (a << 1);

$$\frac{Q_r}{Q_t} = \left(\frac{\exp(U) - U - 1}{\exp(U_s) - U_s - 1}\right)^{\frac{1}{2}} \qquad \text{([2], ch. 4, p. 128)}$$

(iii) In the case of existence of deep levels (due to microcontamination for exemple) and taking into account the above definitions (i) and (ii), one can show that:

a) neglecting the free carriers density n the SCR is narrower than that corresponding to shallow levels of equal density.
b) taking into account n SCR is more extented than the SCR due to shallow levels of equal density.

Il is so pointed out that the length of SCR is very sensitive to deep level presence and then a detailed calculation has to be done particulary for scaled-down microelectronic devices.
It is evident also that in an electronic device a perfect local Electrical Neutrality is not possible even in the so called neutral region.

4 - Electrical Neutrality and Fundamental Electromagnetics

As a final remark, it is worthwhile to notice that the absolute local neutrality is not attained for a fundamental réason: the theory of relativity proves that even the simplest effect of electrical conduction in a material provokes a (very small) departure from the local neutrality: this space charge is in the origin of the self induced magnetic field (Ampere's circuital law) [9.]

5 - References

[1] R. BARON, J.W. MAYER : «Double injection in semiconductors» in Semicond and Semimetals 6 Injection phenomena, R.K. WILLARDSON, A.C. BEER ed. chap. 4, Acad. Press. (1970) (cf. p. 102)
[2] A. MANY, Y. GOLDSTEIN, N.B. GROVER : «Semiconductor Surfaces» Nth Holl. Publ. (1965)
[3] R.A. SMITH : «Semiconductors» Cambr. Univers. Press (1959)
[4] G KAMARINOS these d'Etat ès Sc.: «Etude des Propriétés Electriques d'un semiconducteur soumis à un déséquilibre thermodynamique isotherme» Université de Grenoble (1974)
[5] C.G.B. GARRETT, W.H. BRATTAIN: Phys. Rev. 99, 376 (1955)
[6] R.M. KINGSTON, S.I. NEUSTADTER : J. Appl. Phys. 26, 718 (1955)
[7] Z.T. KUZNICKI : Rev. Phys. Appl. 23, 1313, (1988)
[8] K.G. NICHOLS, E.V. VERNON «Transistor Physics»; Chapman and Hall ed. (1966)
[9] M.A MATZEK, B.K. RUSSEL: Amer. J. of Phys. 36, 905 (1968).

Virtual Device: A New Approach in Microelectronics Device Education

Andy Negoi(1), Alain Guyot(2), Ştefan Bara(1)
and Jacques Zimmermann(1)

(1) LPCS-ENSERG, UMR CNRS 5531, 23 rue des Martyrs, 38016
Grenoble Cedex 1, France
(2) TIMA, 46 Avenue Félix Viallet, 38031 Grenoble, France

1 Introduction

The goal of this paper is to present a tool for simulating generic microelectronic devices using specialized integrated circuits. The Monte Carlo method in three dimensions is used as the most simple and the most physical technique. The virtual device takes the shape of a small scientific calculator provided with analog I/O terminals upon which various apparatus, such as: oscilloscopes, signal analyzers or others in order to perform various measurements, from which physical interpretations can be extracted. With the same specific hardware one can characterize various types of devices (diodes, BJTs, FETs) realized in any material (Si, GaAs...), depending on parameters that can be varied at will. A first demonstrator, a synthetic noise generator exists already.

2 Virtual device: what is inside?

Recent advances in microelectronic technology brought into light new phenomena due to the extreme shrinking of device dimensions. Actually, only few tens or hundreds of electrons participate to the charge transport in transistors, and the device simulators must adapt continually their device models. Also, the device improvement and the deep comprehension of noise phenomena require much simulation time. The software tools are expensive and not always well adapted to the needs of microelectronics education in universities.

As the FPGA technology has come to a great level of complexity, the low cost prototyping of large svstems became possible even in an university framework. Thus, innovative ideas about educational tools can be materialized. The simulation of a whole microelectronic device with dedicated circuits is one of them. The Monte Carlo method has the advantage of tracing the physics of particles at the most basic level, so the connection between solid-state physics, charge transport phenomena

T.J. Mouthaan and C. Salm (eds.), Microelectronics Education, 35-38.

and more complicated device physics is rendered more "easy and natural" than it could be via sophisticated transport equations [1].

The virtual device, a small stand-alone system using FPGAs, implements a Monte Carlo algorithm and is provided with analog terminals (via A/D and D/A converters). Seen by the user, it can be investigated exactly as a real device, using signal analyzers, oscilloscopes and function generators. The difference comes from the possibility to change at will physical parameters in order to observe the effects. One can also change the geometry of the device, its type (uni- or bipolar), the material (Si, GaAs, etc.), so one can recognize the virtual device as a traditional simulator. But these simulators are expensive computer programs, they need computers to work, which are not always available in practical work rooms. Another drawback of these simulators is that they give plots but we cannot always know the simulation method, the physical models are often shadowed. The simplified Monte Carlo algorithm that we used simulates independent particles and overcome these drawbacks. This kind of method is traditionally slow on computers, but a dedicated machine like ours is expected to simulate in real time.

The evolution of each particle in phase space is determined by the Monte Carlo algorithm presented in [2]. Figure 1 shows the structure of the device simulator.

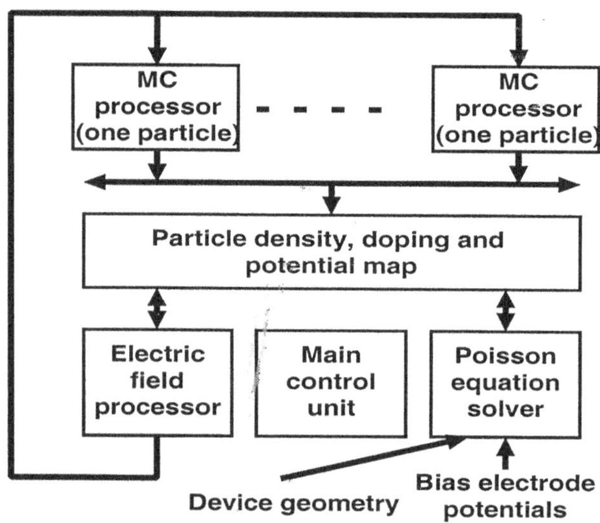

Figure 1: General configuration of an artificial device simulating N particles in parallel.

N processors simulate N charged particles at a time, drawing their positions and velocities. Periodically, with a period to be decided by the user, the position of each particle is taken for updating a particle density map. This information, together with the geometry and doping map (in two dimensions) is used by a finite difference wired-up Poisson equation solver, based on the well-known Gauss-Seidel algorithm applied in an iterative scheme. The potential map is obtained. Each particle processor, using its particle position information, retrieves the potentials in the neighbouring points and calculates the local electric field.

Each part of this system functions in two phases: i) calculate and ii) communicate. The calculating phases of the key blocks (particle processors and Poisson equation solver) have about the same duration and are much longer than the communication phase. A time skewing based method is thus well adapted for the system synchronization, and the virtual device will be very fast compared with a software program.

3 Toying with solid state physics?

The difference between the virtual device and a real one is that by changing physical material parameters and device geometry, the student can "toy with physics", and analyze the effects which are not obvious to observe experimentally due for example to the fact that the imagined cannot be yet fabricated. So far, we have described the most elaborate structure possible. The most simple artificial device would consist in just one unique particle processor. The system in which the processor is embodied then is illustrated in figure 2. With such a system, we

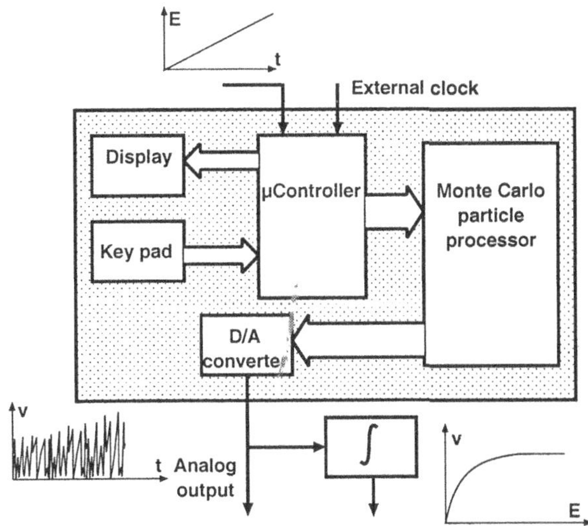

Figure 2: General structure of a virtual device using one particle processor.

can examine various properties of transport of the particle, perform first moment measurements and also second moment measurements (noise) and since there is no limitation on the amplitude of the field applied to the particle, hot carrier simulations are easy and immediate [2]. The processor can be operated at relatively low clock frequency because it must adapt itself to the speed of acquisition of the data at the outside as wished by the user. If we apply a fixed voltage at the input the response at the output is the instantaneous velocity of the particle to be observed as a function of time. If the voltage at the input is slowly swept and the output sent to an integrator and the result sent to a scope time-based with the input, we directly observe the drift velocity as a function of the electric field. In this configuration the various physical parameters, effective mass, phonon energy, doping concentration can be varied by the user and the consequences observed on the scope in real time (analysis of the mobility, saturation velocity, etc). Next, if a signal analyzer is available, the instantaneous velocity can be analyzed (autocorrelation function, noise power spectra) at a fixed field. Then the notion of hot carrier diffusion coefficients due to velocity fluctuations can be physically analyzed too. In the same spirit, the notion of noise factor can be illustrated by putting at the input a small time varying voltage (field) at a given frequency superimposed on the DC voltage. The response to the small time varying field can be extracted from the noise at the output simply using a signal analyzer.

A first demonstrator of the idea has already been achieved. The structure is the one illustrated in fig.2 in which the particle processor has been replaced by a set of 16 looped shift registers multiplexed in sequence. Depending on the number of registers used, the noise transfered to the outside is colored or white. The registers have been implemented in a FPGA. The Monte Carlo processor [3] for one particle and the Poisson equation solver [4] have been implemented on FPGAs and are under test.

4 Conclusion

In conclusion, we have shown how the use of wired-up processors can generate new tools in view of pedagogic applications, for instance in practical works for electrical engineering students. Monte Carlo particle processors are presented here as building blocks for the realisation of such a virtual device.

References

[1] C. Jacoboni and P. Lugli. *The Monte Carlo Method for Semiconductor Device Simulation.* Springer-Verlag, Wien - New York, 1989.

[2] M. Shur. *Physics of Semiconductor Devices*, chapter 1. Prentice Hall, 1990.

[3] A. Negoi, A. Guyot, and J. Zimmermann. A dedicated circuit for charged particles simulation using the Monte Carlo method. In *Proc. ASAP'97*, pages 422 – 431, Zürich, Switzerland, 1997.

[4] A. Negoi and J. Zimmermann. Monte Carlo hardware simulator for electron dynamics in semiconductors. Part II: A parallel approach to the Poisson equation solution. In *Proc. CAS'97*, pages 293 – 296, Sinaia, Romania, 1997.

INTERACTIVE LEARNING ENVIRONMENT FOR THE PRACTICAL TRAINING ON DIGITAL ELECTRONICS

L. RODRIGUEZ-PARDO, M.J. MOURE, M.D.VALDES, E. MANDADO
Departamento de Tecnología Electrónica
University of Vigo
Apartado de Correos Oficial, 36200
Vigo - Spain

1 Abstract

The main objective of Electronic Engineering learning is the development of the student's ability to design, simulate and test electronic circuits to implement real systems. Commercial tools are very flexible and feature a lot of options. Nevertheless, they are oriented towards the professional market, and they provide many features not needed for educational purposes. On the other hand, much time devoted to understanding very complex manuals is required by the students to master these tools, slowing down the learning of fundamental design concepts. The objective of this communication is to present a CAD tool for the design of digital electronic circuits as well as for testing developed circuits. The system, which is called VISCP, is an educational tool for practical training on Digital Systems and is intended to guide students in the practical design of basic digital circuits.

2 Introduction

The fall of prices of personal computers has made their introduction in educational laboratories economically feasible. Their use makes the teaching based on new design techniques such as Schematic Capture Programs or Simulators possible. On the other hand, Virtual Testing Instruments can be developed using additional hardware at low cost, thereby avoiding the need of expensive instrumentation.
The use of commercial tools in educational applications has many inconveniences [1]. The first one is that commercial tools are difficult to manage. The experience of using these tools in laboratory classes demonstrates that students need to spend a considerable amount of time in managing the program interface, thereby causing a reduction of the time that they dedicate to the circuits design. On the other hand, commercial tools are usually oriented towards professional applications and then, they do not provide help information about how to accomplish a design. Their help information is reduced to the program options. Also, there is not a commercial system that allows the design, simulation and verification of electronic circuits within a unique environment. These drawbacks have been the motivating factors that lead to the development of an own

T.J. Mouthaan and C. Salm (eds.), Microelectronics Education, 39-42.

system, called VISCP (Virtual Instrumentation and Schematic Capture Program), instead of using commercial tools for the practical teaching of electronics [2].
The principal aim of VISCP is to provide students with a simple and easy to use Interactive Design Environment instead of developing a system with the best performance. VISCP comprises of two main modules: a Schematic Capture Program (SCP) for circuit design, with a Netlist Generation Tool that interfaces with commercial simulators, and a Virtual Instrumentation (VI) module for test purpose. The block diagram of the system is shown in Fig.1.

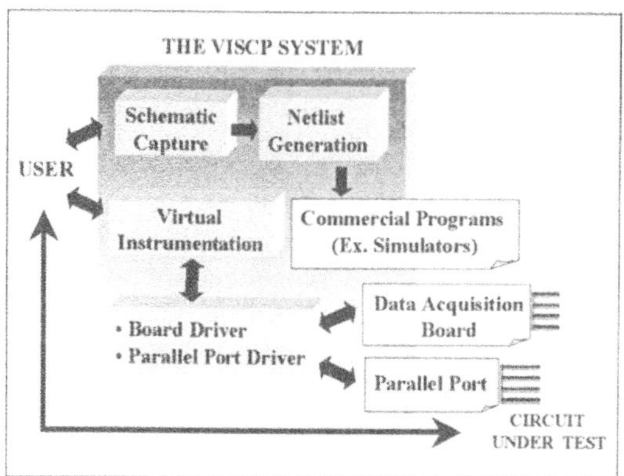

Fig. 1. Block Diagram of the VISCP System

3 Schematic Capture Program

Recently, the development of professional CAD tools has improved the design task in Electronics. At present, its use is indispensable in the case of complicated designs, so the laboratory teaching based on these new design techniques is mandatory. The students can design prototypes of systems with the help of a Schematic Capture Tool. Then, they can simulate its operation and ensure that the performance of the prototype is the desired one. If the results of the simulation are the expected ones, they can proceed to assemble and test the real circuit. If not, they must change their design and simulate its operation again. The simulation helps students to test and refine its designs before assembling the circuit with real devices.
The SCP of VISCP provides students with a useful and easy to use environment for the design of digital circuits. Its friendly and simple interface allows students to master the program performance in a short space of time. Help on line is always available, and the component libraries have been improved in order to provide information to the student about the operation mode of the devices. Also, the SCP has been designed to help students in the design task by preventing typical mistakes in designs (for example, connections that can damage the devices in a real assembly are not permitted), and allowing them to link their designs with commercial simulators like PSpice and with PCB design programs, without the need of managing their configuration and input parameters.

4 Virtual Instrumentation

The testing of prototype systems is one of the main steps in the design of systems because its results can validate the operation of the systems or, on the contrary, can lead to the modification of some design aspects. Therefore, it is important that the students obtain the basis of the verification of systems at the same time as they learn to design them [3].

How to test a system is a skill that can only be completely accomplished by means of experience and practice. The test strategy varies with the technology used, the final application of the system, the available points of measure, etc. Due to this, verification is always a difficult matter for students and, therefore, the learning of verification strategies must be one of the fundamental objectives of the laboratory classes [4].

Nowadays, the cost of the test process, in effort and time, is reduced by the use of professional measurement instruments. But the introduction of these instruments in laboratory classes in the first years of the degree presents some drawbacks. Firstly, commercial instruments provide many features that are not required for educational purposes and consequently, the inversion in instrumentation is not justified by the final use of the equipment. Secondly, the managing of commercial instruments is difficult because of their complex functions oriented toward the professional market. For this reason the instruments needed for educational purposes must be simple and interactive and its functionality must be adapted to the knowledge of the students. The Virtual Instrumentation technique provides an easy access to achieve these purposes.

The virtual instrumentation technique is based on combining a computer with a simple and general data acquisition system in order to define a specific measuring instrument. The main advantages of this approach are the following:

1) *More efficient use of hardware resources:* The hardware is common to any instrument so the use of resources is more efficient. Using the same hardware, software modules can be developed in order to generate different instruments like an oscilloscope, a logic analyzer or a multimeter.

2) *Adaptation of the control panel to the student's degree of knowledge*: The software determines the operation mode. In this way, the control panel can be adapted to the application, according to the requirements of the level of knowledge of the student. Besides, the students find the computer environment more familiar and then more manageable than the control panel of any traditional instrument.

3) *Several applications at the same time*: Using actual multitasking operating systems (for example Windows95 for PCs), screens of different active programs can be represented simultaneously. In this way, the acquired data can be compared, for example, with simulation results easily. Moreover, data files can be shared between several applications and the students can analyze their results with datasheet programs, mathematical tools, etc.

4.1. VIRTUAL INSTRUMENTS OF VISCP

Nowadays, VISCP works over PC platform and any data acquisition board can be used as interface to the external world. Only a small configuration must be done for each acquisition card. Another developed version of VISCP is oriented to distance learning and uses the parallel port of the PC as interface to external signals. In this way, only a reduced hardware must be added to the PC in order to operate like a virtual instrument.

Up to now, two virtual instruments for Digital System verification have been developed: a Voltmeter/Logic Probe and a Logic Analyzer. The Voltmeter/Logic-Probe is suitable for the test of combinational circuits while the Logic Analyzer facilitates the comprehension and test of sequential circuits. The instruments are generated from the screen of the Schematics Capture and the student can associate the external test point and the corresponding node of the schematic. In this way, the results or waveforms are directly visualized on the schematic, thereby facilitating the test.
In Fig. 2 the schematic of a circuit and its test through a Logic Analyzer operating as a State Analyzer and Volmeters/Logic Probes is represented.

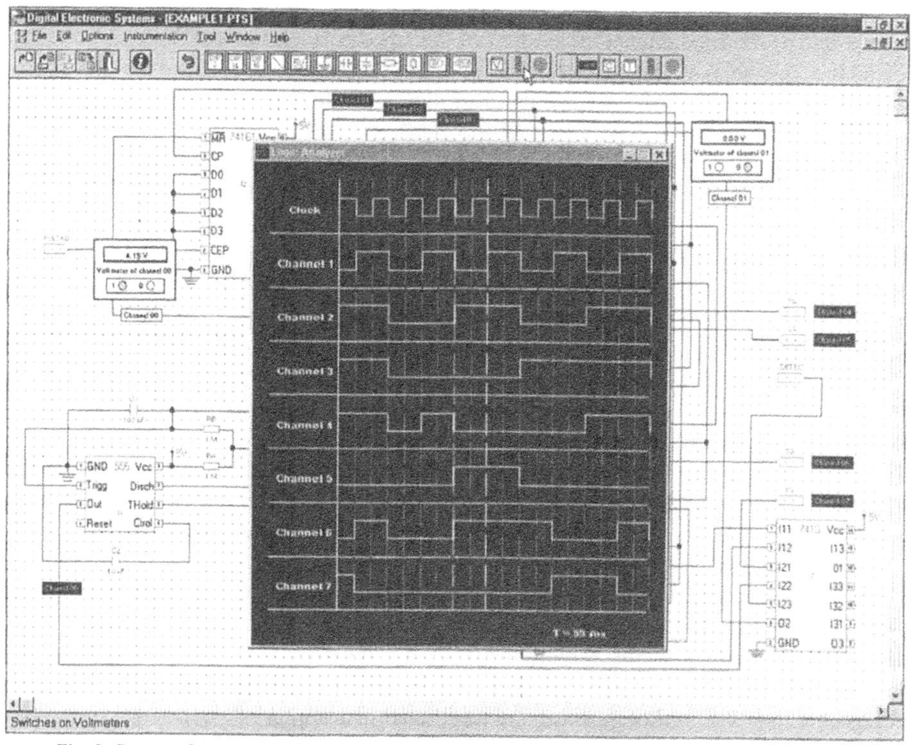

Fig. 2. Screen of the VISCP System. Schematic of a circuit and its test through a Logic Analyzer.

5 References

[1] P. Galvis, "Ingeniería del Software Educativo", Bogotá: Publicaciones de la Universidad de los Andes, 1992.

[2] E. Mandado, D. Valdés, M.J. Moure and L. Rodríguez, "VISCP: a PC-based system for practical training of Digital Electronic Circuits Design and Verification", ALT-C 96 Association for Learning Technology (third annual conference), Glasgow, Scotland, UK, 1996.

[3] M.J.Moure, L.Rodríguez-Pardo, M.D.Valdés, E.Mandado "An Interactive Environment for Digital Electronics Learning and Design", Proceedings of ED-MEDIA 97 &ED-TELECOM 97 (World Conference on Educational Multimedia and Hypermedia & on Educational Telecommunications), AACE, Calgary, Canada, 1997.

[4] M.D. Valdés, M.J. Moure, L. Rodríguez and A. del Río, "Interactive Practical Teaching of Digital Circuits Design by means of Field Programmable Gate Arrays", Computer Aided Learning and Instruction in Science and Engineering, Springer-Verlag (Eds), 1996, pp. 408-414.

AN INTERACTIVE ELECTRONICS COURSE USING PSPICE

JULIO J. GONZÁLEZ[†], ENRIQUE MANDADO[‡]
[†] *Department of Electrical Engineering*
State University of New York at New Paltz, USA
gonzalj@eelab.newpaltz.edu
[‡] *Department. of Electronic Technology*
University of Vigo, Spain
emandado@uvigo.es

1 Abstract

This paper describes a computer-assisted interactive methodology for teaching Electronics.

2 Description of Interactive Method

The interactive approach requires a computer assisted classroom (usually the Electronics laboratory) equipped with one computer per group of students (usually a maximum number of three). The teaching method, which consists of three phases, will be explained with the aid of Figure 1.

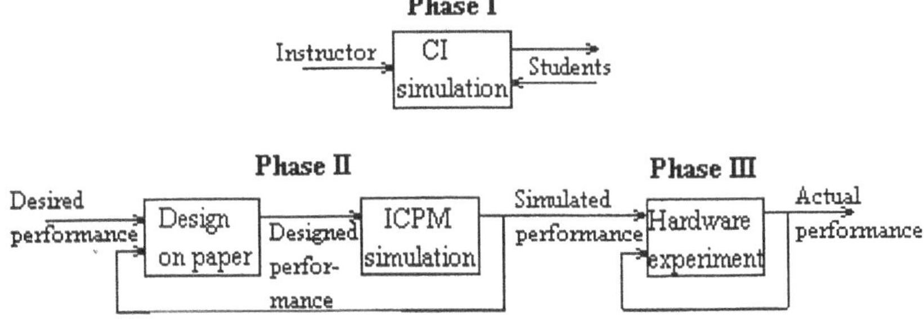

Figure 1

The three phases will be illustrated using "Transistor Biasing" as an example topic.

T.J. Mouthaan and C. Salm (eds.), Microelectronics Education, 43-45.

Phase I (In class): The Instructor explains the topic on the board. She/he discusses the effects of parameter variation on biasing stability and the stability improvement provided by feedback. She/he then calculates the biasing component values for a given tolerated biasing change. Then the students proceed to input these values in a customized interactive (CI) Pspice program, and verify the instructor's design. Finally, they change one biasing value at a time and discuss the corresponding simulation results among themselves and with the instructor.

Phase II (Off class): As a "homework" assignment, the students are asked to produce several designs for different situations (like different parameter variations or different tolerated biasing changes), and to verify their designs using an Initially-Customized Progressively Modifiable (ICPM) Pspice program. The objective of ICPM programs will be explained in Section 3. Students are also requested to prepare a brief report consisting of their design, the corresponding Pspice print outs, and their interpretation of the simulation results. This Pre-Lab. work is a necessary requisite to allow students to perform the Laboratory Experiments in Phase III.

Phase III (In the Laboratory): In the Laboratory Session, the students verify experimentally the results in Phase II on an electronic test-board.

As it can be seen, in this teaching method, theory, simulation and experimentation become integral parts of a complete process. The interaction in Phase I adds to the teaching dynamics and prevents the student from loosing concentration, since she/he is no longer a passive element of the class. The work in Phase II serves the double purpose of re-affirming the concepts acquired in Phase I and providing the student with an expectation for the experimental result that should be obtained in Phase III. Finally, since Phase III is not unrelated to the theoretical class (as usually in current practice), it logically completes the teaching cycle. The described method fits the hands-on-expertise year 2000 criteria of the American Accreditation Board of Engineering and Technology (ABET).

3 Implementation of Interactive Method

Both Phases I and II require the use of customized Pspice programs. In Phase I, the reason for customization is obvious: the instructor needs a previously written program which is ready for interacting with students (CI program).

In Phase II, the Pspice programs are initially customized to the specific experiments of the first laboratory sessions, and become progressively modifiable by the student throughout the remaining laboratory sessions (ICPM programs). The need for ICPM programs derives from the facts that (I) programming details divert the students' attention from the primary goal of fully understanding the circuit under study, and (II) it is impossible to obtain profit from a given simulator without having an excellent grasp of the corresponding theory. In the ICPM educational system, the demand for programming knowledge is increased sequentially, at a pace compatible with the student's assimilation of Electronics theory. The benefits of the ICPM system are discussed in detail in [1].

Both the CI and the ICPM programs consist of constant code lines (i.e., they cannot be changed by the user) and variable code lines (i.e., they can be changed by the user). In a CI program, the only variable lines are the lines where the user (either the instructor or the student) inputs the corresponding design data. In the ICPM system, subsequent programs incorporate progressively an increasing number of variable code lines. These lines allow the student to input not only design data, but also different circuit parameters, such as frequency, temperature, as well as circuit components of different characteristics.

The constant and variable code lines are implemented using the Excel capability of locking and unlocking cells. The constant code is placed in locked cells, and the variable code in unlocked cells. The locked cells can be unlocked only in possession of the instructor's password.

The format of the Excel file is not suitable for being run by Pspice. This small inconvenience is solved by (I) saving the Excel file as a text file, and (II) changing the Excel extension ".xls" into the Pspice extension ".cir".

4 Discussion

The CI and ICPM programs have been developed by SUNY New Paltz students working on a AMP/C-STEP grant funded by the National Science Foundation (NSF) and the New York State Department of Education. The described teaching methodology is currently being implemented at the Department of Electrical Engineering of SUNY at New Paltz, USA, for teaching the Electronics I course.

Preliminary results are excellent. The conjunction of theory, simulation and practice in this teaching methodology solidly contributes to a deep understanding of Electronics by the student. Critical aspects which deserve particular attention to guarantee success of this educational system are (I) a dynamic teaching pace to cover theoretically all the laboratory experiments, and (II) a progressive increase in the amount of Pspice code that can be modified by the student.

5 References

[1] Julio J. González, "Toward an Optimized Computer Assisted Electronics Laboratory," Proceedings of the 1997 IEEE Computer Society International Conference on Microelectronic Systems Education, Arlington, Virginia, pp. 55-56, July 21-23 1997.

MICROELECTRONICS TEACHING – THE GLASGOW APPROACH

I.G. THAYNE
Dept of Electronics and Electrical Engineering
University of Glasgow, Glasgow, G12 8LT, Scotland, UK

1 Introduction

This paper describes a final year undergraduate course in Microelectronics whose aim is to introduce the student to the methodology of analogue integrated circuit design by exposing them to the complete design cycle. On average, 25 students attend this course.

The course is taught in ten blocks of four hours, with no more than one hour of each block devoted to lecturing - the rest of the time is used for practical design exercises. The teaching material is complemented by seminars on mixed signal design from the technical director of Wolfson Microelectronics in Edinburgh, and a visit to the Motorola microcontroller design centre in East Kilbride.

The circuits designed by the students on the course are fabricated using an in-house 1μm GaAs MESFET process to minimise turn-around time. Timetabling requires that the circuits be produced in eight weeks to enable the students to complete the testing phase of the course.

Given the time constraints and the underpinning aim of exposing the students to *all* the phases of custom circuit design (design, layout, fabrication and testing), circuit complexity and functionality is minimised.

The following section outlines the course details.

2 Course Details

The course can best be summarised under the five areas comprising the "complete design cycle" shown in Figure 1.

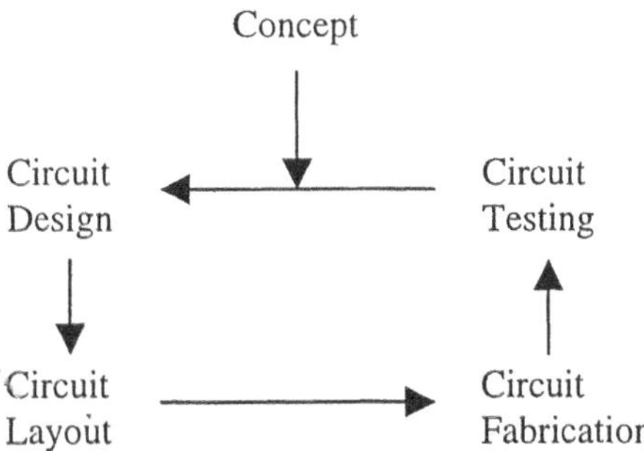

Figure 1 – The Complete Design Cycle of the Glasgow Microelectronics Course

T.J. Mouthaan and C. Salm (eds.), Microelectronics Education, 47-50.

2.1 CONCEPT

The circuit studied in recent years has been a self-biased cascoded inverting amplifier consisting of eight MESFETs shown in Figure 2. The lecture/lab schedule summarised in Table 1 leads the student to the point where they can design such circuits and lay them out for fabrication. Testing is performed at the end of the 2nd term with a gap of eight weeks to permit circuit fabrication.

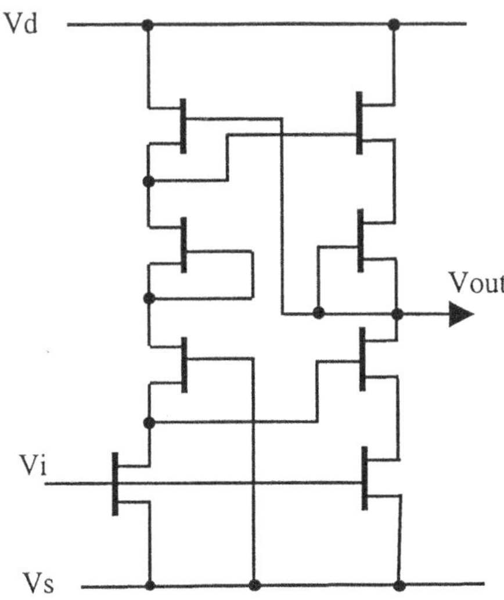

Figure 2 – Self-biased Cascoded Inverting Amplifier

Table 1 – Summary of Course Lectures and Design Exercises

Week	Lecture	Design Exercise
1	Introduction to MESFETs	MathCad – GaAs MESFET Operation
2	MESFETs at DC and AC	Construction and testing of simple hybrid FET circuits
3	Simpler Inverting Amplifier	pSPICE of Simple Inverting Amplifier
4	Cascoding, what, why and how	PSPICE of Cascode Inverting Amplifier
5	Tutorial	Group Study of Other Building Blocks (current mirrors and transconductors)
6	Self-biasing circuits for cascoding	PSPICE of self-biased Cascode Inverting Amplifier
7	Introduction to Layout	Circuit Layout
8	Further Layout Issues and Decoupling	Circuit Layout
9	Mixed Signal Design Seminar	
10	Mixed Signal Design Seminar	
13	Visit to Motorola Microcontroller Design Centre	
18	Circuit Testing	
19	Circuit Testing and Outcomes	

2.2 CIRCUIT DESIGN

The student firstly investigates the performance of a simple source-follower circuit using pSPICE™ to gain familiarity with the issues associated with designing using a GaAs MESFET process. The MESFET is described in pSPICE using a Parker-Skellern model[1] extracted using FETFIT[2].

Next, optimisation of the performance of a number of inverting amplifier topologies of increasing complexity is undertaken.

A simple inverting amplifier is first studied to give the student further exposure to the constraints of designing using a GaAs MESFET process. This exercise leads to the optimisation of a cascoded driver inverting amplifier which is a stepping stone to the final self-biased cascoded inverting amplifier topology. Each student chooses a parameter to optimise, eg gain, bandwidth, minimum DC power consumption and designs the amplifiers accordingly.

2.3 CIRCUIT LAYOUT

The student lays out the source follower, simple inverting amplifier and self-biased cascoded inverting amplifier circuits using Wavemaker™, adhering to design rules discussed in the lectures. Prior to fabrication, the layouts are run through the Cadence™ Design Rule Checker using a technology file description of the Glasgow process. Figure 3 shows a typical layout generated at the completion of this phase of the work. The chip area is 4 mm x 4 mm including a range of generic process control monitors and alignment mark structures which are dropped into the layout after its submission by the student for fabrication.

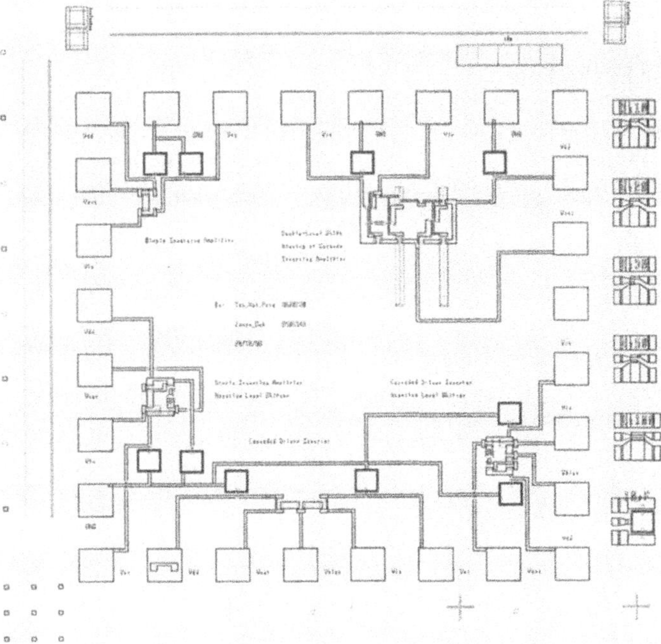

Figure 3 – Typical Layout Submitted for Fabrication

2.4 CIRCUIT FABRICATION

The circuits are fabricated by experienced researchers and technical staff using an in-house 1μm GaAs MESFET process. This is a modification of a standard 0.2 μm GaAs MESFET process which has been used to produce a range of microwave and millimetre-wave integrated circuits[3]. The process is based on an epitaxial GaAs MESFET substrate with all levels defined using direct write electron beam lithography realised using a Leica-Cambridge EBPG-5 Beamwriter[4]. Direct write is used to minimise the fabrication time as the need for the generation of optical reticles is obviated. Although they do not personally perform the circuit fabrication, the students are invited to spend time in the cleanrooms during this phase to witness the progress of the circuits.

2.5 CIRCUIT TESTING

The completed wafers are diced and bonded in ceramic PLLC packages which drop into burn-in sockets on a testing PCB. Provision is made on the PCB for the addition of off-chip decoupling capacitors should they be required. The aim of the testing phase is to complete the design cycle by comparing the circuit performance with the original pSPICE™ predictions and explaining any deviations. Typical discrepancies observed in recent years include process variations leading to modifications in the performance of the MESFETs and parasitic capacitance and inductance associated with layout issues. Whilst such effects are clearly undesirable, their occurrance in a teaching environment leads to a deeper understanding of "non-ideal" issues often overlooked in courses with a less practical emphasis.

3 Course Evaluation

Given its practical nature, 80% of the final mark of this course is based on an individual report describing the work undertaken in the design, layout and testing of the circuits with the remaining 20% allocated in a written examination.

4 References

[1] A.E. Parker, D.J. Skellern, "Improved MESFET Characterisation for Analog Circuit Design and Analysis", GaAs IC Symposium 1992

[2] D.R. Webster et al, "Problems in Designing FET MMICs with Low Distortion", IEE Colloquium on Modelling, Design and Application of MMICs, June 1994

[3] I.G. Thayne et al, "Front-end Building Blocks using the Glasgow 0.2 mm GaAs MESFET Process", IEE Colloquium on RF and Microwave Circuits for Commercial Applications, February 1997

[4] N.I. Cameron et al. "Selectively Dry Gate Recess Etched GaAs MESFETs, HEMTs and MMICs", J. Vac. Sci. Techol. B, 11, 6, 2244, 1993

Session P2

EMERGING FIELDS/ TECHNOLOGY

EDUCATION CONCEPTS ON III-V BASED MICROSYSTEMS

G. JACQUEMOD, F. GAFFIOT, P. ROJO-ROMEO
J.L. LECLERCQ, X. LETARTRE, P. VIKTOROVITCH

Ecole Centrale de Lyon, LEAME UMR CNRS 5512
36 Avenue Guy de Collongue, B.P. 163
F-69 131 Ecully Cedex, FRANCE

1 Abstract

A highly selective selective and tunable optical filter at 1.55 μm using a Fabry-Perot resonator with micromachined InP-air gap distributed Bragg reflectors is developed in our laboratory. In order to optimize the global characteristics of the MOEMS, we have developed an opto-electro-mechanical model. The behavioral model, written in VHDL-AMS-like, allows to simulate each part of this device (electrostatic actuation, mechanical response, optical filtering) separately. Moreover, an integrated simulation, taking into account the interactions between the different parts of the model, is also possible. This model has been implemented into a classical electronic simulator framework, and thus allows the simulation of an entire WDM communication system. The simulated network (STARNET-II-like) consists of four nodes, each one includes an optical filter for the control and a tunable one for the data.

2 Introduction

The department of electronics proposes "classical" courses on microelectronics (heterostructures, analogue circuits and systems design, VLSI architecture, design techniques of modern electronic systems, ...). Theoretical backgrounds are associated with practical training on microelectronics techniques (clean room main steps for microelectronics devices manufacturing, electrical characterisation of III-V heterostructures, ...) and CAD tools (electrical simulation for analogue circuits, full custom design tools, hardware description languages).
For a few years, specific training on microsystems has appeared. First of all, the theoretical principles of bulk and surface micromachining for III-V semiconductor structures are pointed out. Then, practical activities allow students to process free standing beams, using clean room facilities. These devices are designed only by manual parameter targeting based on the analysis of behavioural simulation results. The trends consist on integrating the behavioural model and the technological process in a CAD framework in order to allow parameter extraction and optimisation.

T.J. Mouthaan and C. Salm (eds.), Microelectronics Education, 53-56.

Our experience has led to new concepts on III-V based microsystems, firstly MOEMS (micro-opto-electro-mechanical system) : due to transfer of the processing technology developed for microoptoelectronics to the field of micromechanics with the monolithic integration of electro-optical functions.
This paper presents the design and the fabrication of a vertical micromachined InP-air gap based Fabry-Perot filter, continuously tunable via electrostatic actuation of the cavity air gap, which meets the WDM system requirements. A behavioral model of this device, written using VHDL-AMS-like Hardware Description Language HDL-A, has been developed. The behavioral modeling approach is used for the co-simulation of network access control with physical layer photonics and electronics for a WDM optical communication network

3 MOEMS design

The filters comprise of a Fabry-Perot air gap resonant cavity formed between two high reflectance Bragg mirrors based on InP/air-gap pairs. Due to the high index contrast between the air and the InP quarter-wave layers, reflectivity as high as 99.9 % is achieved with only 2.5 pairs. Taking into account electrical, mechanical and optical parameters a filter design has been realized (fig. 1).
All reflectors and cavity air gaps can be easily fabricated by selective micromachining of InGaAs sacrificial layers. The top mirror is p-doped, the cavity is non-intentionally doped and the bottom mirror is n-doped to form a p-i-n junction, which can be reversely biased. The electrostatic tuning voltage is applied to the layers adjacent to the Fabry-Perot cavity, thus reducing its thickness and tuning the resonant wavelength. The half-wave thick resonant cavity has been designed for a center wavelength of 1570 nm (with no tuning bias), in order to operate in the range 1540-1570 nm.
Surface micromachining technology is used to fabricate the filter : it implies a first step of non selective vertical patterning by Reactive Ion Etching (RIE) with CH_4:H_2 ; in a second step, the suspended layers are freed by selective lateral wet underetching of the InGaAs sacrificial layers with HF:H_2O_2:H_2O solution. Smooth optical graded non-sticking multi air gap InP paddle-shaped bridges were fabricated (fig 1) : the central (30x30 μm^2) platforms are suspended by two, 5 μm wide, arms.

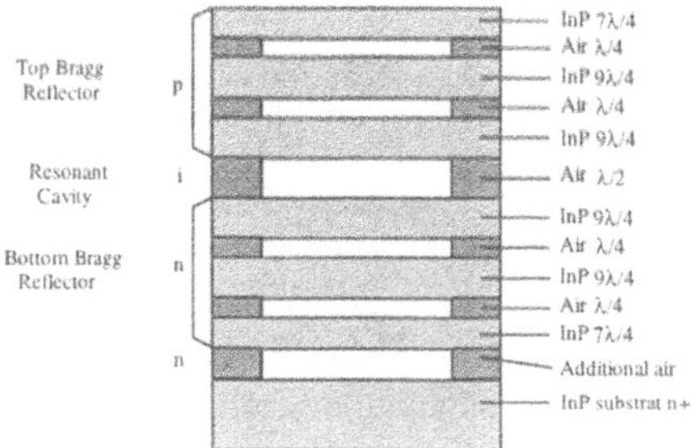

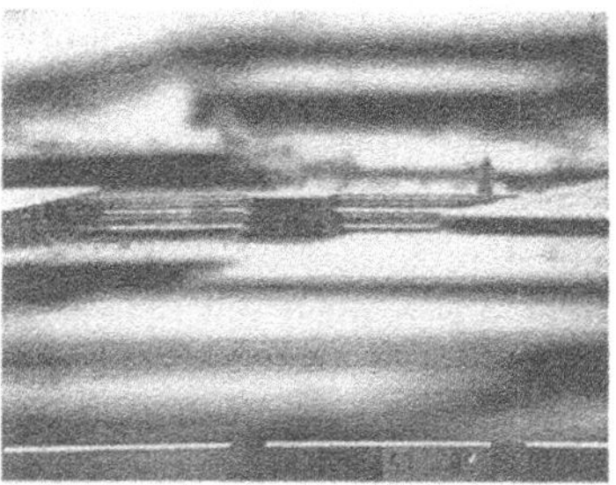

Figure 1 : Schematic structure (left) and SEM micrograph (right) of the MOEMS

4 Modeling of the MOEMS using HDL-A

The behavioral model of the optical tunable filter is shown in figure 2. Electrically, the MOEMS behaves like a reversed biased pin diode. The voltage induces an electrostatic force bringing the beams (next to the cavity) closer together, thus reducing the thickness and the resonant frequency of the cavity.

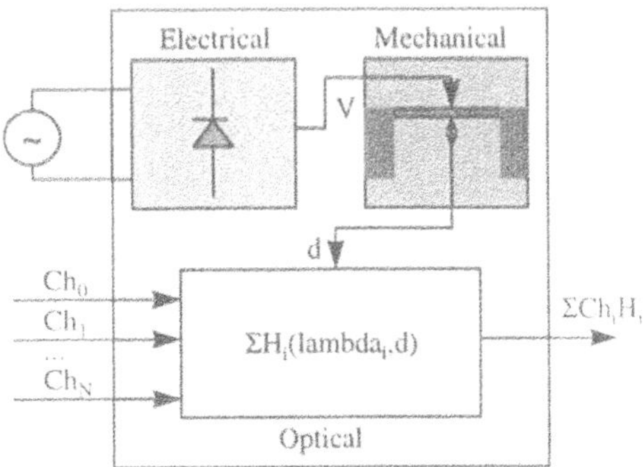

Figure 2 : Opto-electro-mechanical behavioral model of the MOEMS

Here, we have neglected the electrostatic interactions. Consequently, the cavity thickness w remains constant in electrostatic force model, $w_0=\lambda/2=0.775\mu m$, during the actuation. The clamped-clamped beam has several mechanical resonance modes. However, considering only the first mode gives a good approximation of the bending (error < 2%). Thus, the dynamics of the beam resemble those of a simple harmonic oscillator characterized by :

$$M\frac{d^2x}{dt^2} + \gamma\frac{dx}{dt} + kx = \varepsilon\frac{SV_D^2}{2w^2}$$

where t is time, x is the displacement, M the mass and S the area of the beam. γ is the damping coefficient, k the spring constant, V_D the applied voltage, ε the air permittivity.

The optical behavior of MOEMS is determined by transfer matrix method. The global transfer matrix [T] of a MOEMS, including N successive layers, is a product of transfer matrices $[T_k]$ of each individual layers :

$$[T] = \begin{pmatrix} T_{11} & T_{12} \\ T_{21} & T_{22} \end{pmatrix} = [T_1][T_2]...[T_N]$$

where T_{ij} are known as the transmission coefficients of the structure.

Figure 3.a shows the comparison between experimental and simulated results in term of absorption for a zero bias and figure 3.b illustrated the transient response for an optimal multi-step actuation.

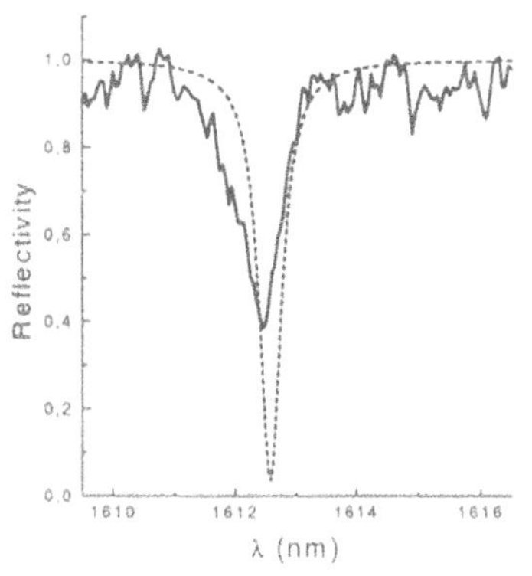

a - Experimental and simulated results

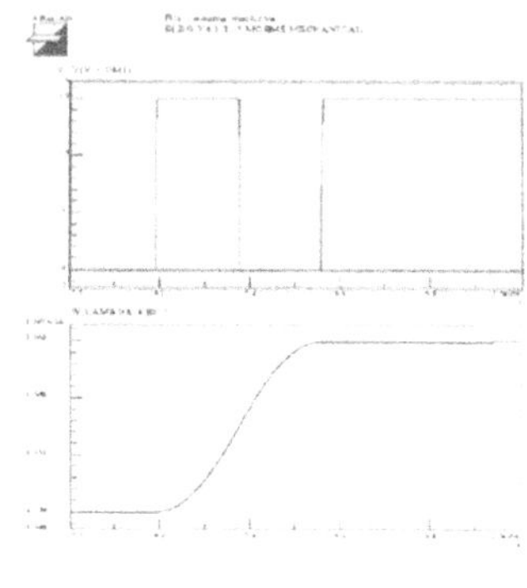

b - Optimal transient response

figure 3 : Simulation results

This model has been implemented into a classical electronic simulation framework. The simulated network (STARNET-II-like) consists of four nodes, each one includes an optical filter for the control and a tunable one for the data. The tuning delay requirements are crucial for time division multiplexed WDM networks. Figure 4.a gives the variation of the tuning delay in function of the beam length. The cross-talk penalty in function of channel spacing is illustrated by figure 4.b.

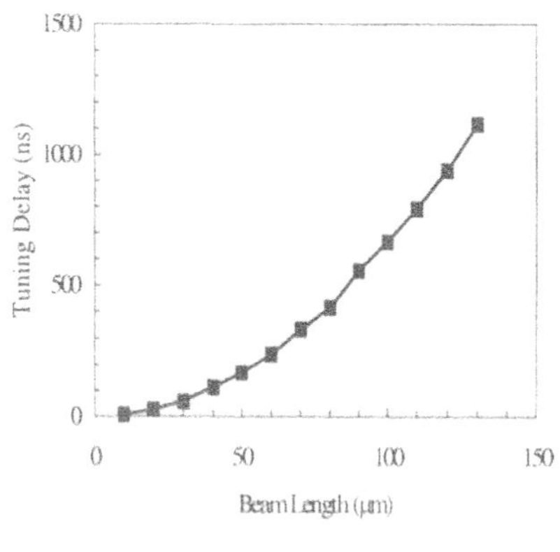

a - Tuning delay

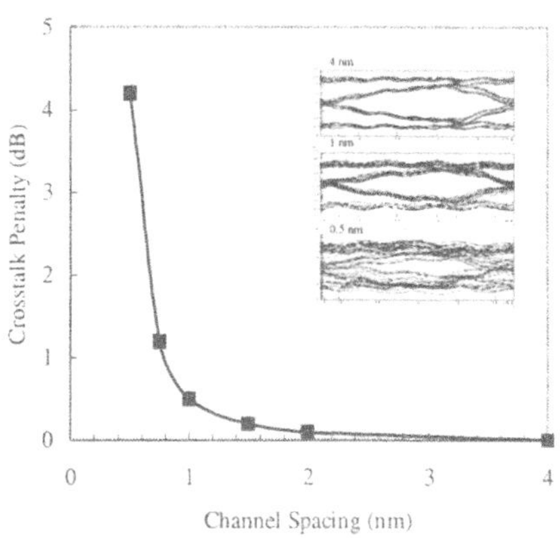

b - Cross-talk penalty

Figure 4 : Simulated tuning delay and cross-talk penalty

MICROSYSTEMS EDUCATION WITHOUT USING THE CLEAN-ROOM ENVIRONMENT : STUDY OF AN ACCELEROMETER REALIZED USING PRINTED CIRCUIT TECHNOLOGY.

Gilles AMENDOLA, Lionel BABADJIAN
groupe ESIEE, Chambre de Commerce et d'Industrie de Paris
cité Descartes, BP 99, 2 Bd Blaise-Pascal, F 93162 Noisy-le-Grand, FRANCE.
email : amendolg@esiee.fr, babadjil@esiee.fr

Abstract.

Working on a capacitive accelerometer allows students to learn about microsystems with a very simple technology. Starting from the specifications of the sensor, which have previously been computed (different for each group), students calculate its dimensions. When it is assembled, measurements are done in several conditions (continued accelerations and pulse excitation...). Specific electronic circuits are developed to get different oscillograms (JFET charge amplifier, variable frequency oscillator, controlled gain amplifier). From theses experimental results we can get the values of parameters in the theoretical model. Different kinds of measurements allow cross checking to test the coherence of the results.
Finally, more specific integrated circuits for microsystems are simulated with tools using behavioural or electrical models (MATRIXx, ELDO). These circuits are based on the principle of modulation and demodulation of the signal.

1 Introduction

Projects has always taken a large place in education at ESIEE. This example is a case study for students specializing in electronics and microelectronics. The sensor described here is a capacitive accelerometer. Beginning with the specifications of the sensor, students calculate its dimensions, build it and make measurements for characterization. For the tests they have to develop several kinds of dedicated electronic circuits.

T.J. Mouthaan and C. Salm (eds.), Microelectronics Education, 57-60.

The first paragraph is on the pedagogical aims of this course. Then we describe the sensor and the physical phenomena that the specifications of the accelerometer depend on. Finally we speak about some electronic circuits associated to the sensor. These are built in electronic models or they are simulated using tools for integrated circuits.

2 Pedagogical aims.

Microsystems education often means using a clean room. This is not always easy, because of several problems such as cost, equipment availability and the duration of practical training. Our example allows students to work as microsystem designers. They have to use the basic knowledge they got earlier (physical, measurements and integrated circuits). The specifications of the sensor are defined by three physical parameters. Each students group has its own specifications for the sensor (that have previously been computed).

The first task is to calculate the dimensions of the sensor. When it is finished capacitance measurements are done and oscillograms are taken. From these experimental results we can get the values of parameters in the theoretical model. Different kinds of measurements allow cross-checking to test the coherence of circuit the results.

Several kinds of electronic circuits are built to get oscillograms. Among them there are a JFET charge amplifier and an Operational amplifier amplitude modulation circuit. An integrated circuit using a synchronous modulation and demodulation principle is simulated through analog CAD tools.

3 Sensor description.

Our sensor is an accelerometer with a capacitive detection [1, 2]. It is made of a brass cantilever ended by a plate which is in front of an electrode (figure 1). This device is realized by photolithography through printed circuit technology. The plate and the electrode make up a variable capacitance versus acceleration. Its sensitivity is about one picofarad for an acceleration of 1g.

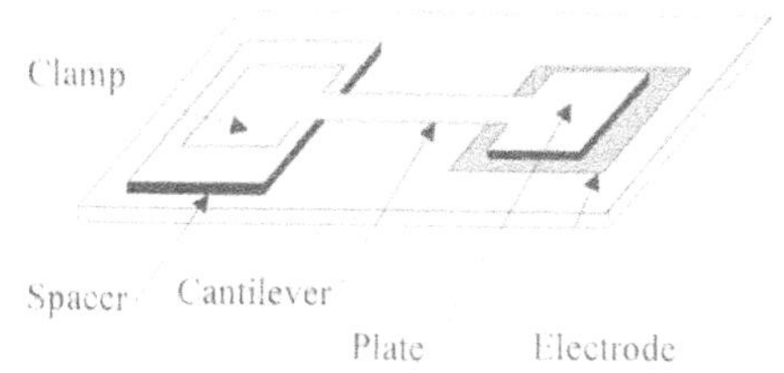

Figure 1

The sensibility is defined here as the difference between the capacitance values when the sensor is working under accelerations of +1g and 0g. As the sensor is similar to an equivalent mass-spring system, its resonant frequency is given by $\omega = \sqrt{k/M}$.

The cantilever's equivalent spring constant k is calculated from the linear elasticity equations. Finally the sticking voltage is the value of the voltage - when it is applied to the variable capacity - which brings the two electrodes in contact.

From several kinds of measurements related to the sensor specification we are able to find some parameters of the theoretical model. For example, we get the equivalent spring constant k and then we can calculate the material Young modulus.

4 Electronics.

4-1 Charge amplifier using a JFET.

A simple solution, using a JFET, gives an output voltage which is proportional to the capacitance variations of the sensor (figure 2a : the accelerometer is represented in the schematic by the capacitance C_0+dC). This schematic has two drawbacks :
- continued accelerations are not detected
- the sensibility of both sensor and amplifier depends on parasitic capacitances

4-2 Amplitude modulation circuit.

The gain of the amplifier (figure 2b) depends on the capacitance value of the sensor, for larger frequencies than $\frac{1}{2\pi R_p C_1}$ we have

$$\frac{V_s}{V_e} = \frac{C_0 + dC}{C_1} \qquad (1)$$

This is able to detect continued accelerations and is insensible to parasitic capacitances.

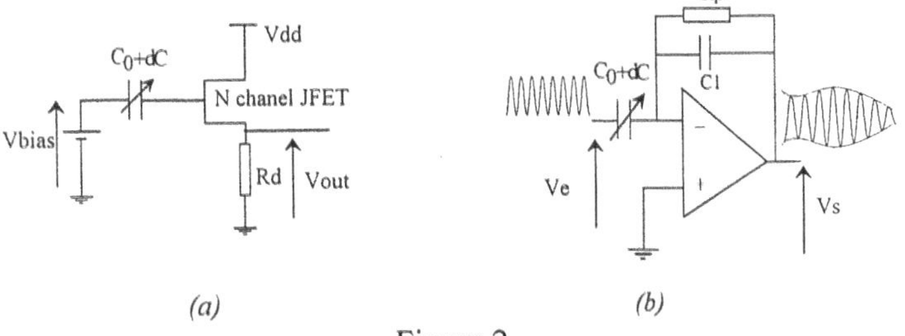

Figure 2

4-3 Simulation of an interfacial circuit CMOS technology compatible.

For actual microsystems, the relative capacitance variations can be as small as 10^{-6} .
In some conditions, the circuit of the figure 3 allows to subtract the influence of the nominal capacity from the output signal. This circuit does synchronous modulation and demodulation and so eliminates the influence of Flicker noise in the amplifier bloc [3] .

$$VS = G \times Vh.[\frac{dC}{C_1} + \frac{C_0 - C_r}{C_1}] \quad \textbf{(2)}$$

G : voltage gain of the amplifier
Vh : voltage of the signal applied to the variable capacitance (the sensor)

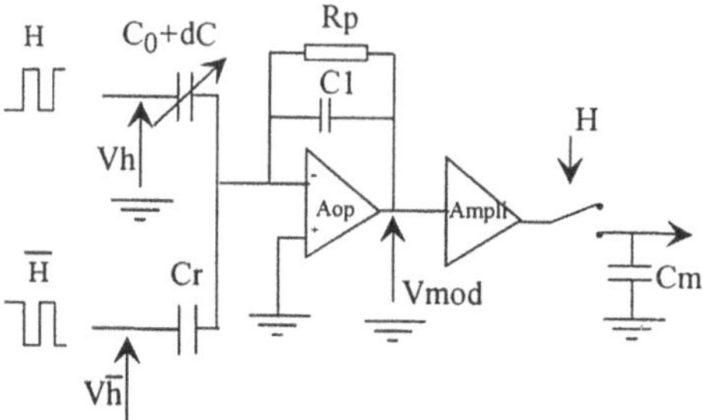

Figure 3

We used the ELDO simulator and behavioural models for the amplifiers. Students look at possible delay in demodulation clocks, capacitance mismatching and the influences of amplifiers defects.

Conclusion

This case study is a synthesis of a large number of fields of knowledge. In this education unit, using a more simple technology, the problems of using the clean-room has been shaped. The course inculcates the basics of microsystems design. Students are faced with a real case to study and observe with satisfaction that theoretical principles can be practically verified.

references.

[1] S.C. Terry, "A miniature accelerometer with built-in damping", Trans.IEEE Solid-State Sensors and Actuators Workshop, Hilton Head Island, sc, USA, June 6-9, 1988, pp. 114-116.

[2] E. Peeters, S. Vergote, B. Puers, W. Sansen, "A highly symmetrical capacitive micro-accelerometer with single degree-of-freedom response", Transducers'91 Digest of Technical Papers, pp. 97-100.

[3] B. de Geeter, O Nys, J-P.Bardyn, "A wide temperature micropower sensor interface circuit" in Proc of 22nd ESSCIRC Neuchâtel Switzland pp136-139, 1996.

EDUCATION IN MICROSYSTEM DESIGN AND REALIZATION

M. HUSAK
Department of Microelectronics, Faculty of Electrical Engineering,
Czech Technical University in Prague
Technická 2, CZ-166 27 Prague 6, Czech Republic
Tel.: +420-2-24352267, fax: +420-2-24310792,
e-mail: husak@feld.cvut.cz

1 Abstract

In the paper there is presented information about ways and results of both education as well as research in the area of Microsystems at the Czech Technical University in Prague. Laboratory equipment, hardware and software tools utilized for microsystem design and testing are described.

The CAD laboratory offers a variety of approaches to education of advanced undergraduate and postgraduate students. The laboratory can serve as a training center for engineers from the industry as well. The main results of research carried out at the department are presented. The research work is oriented towards the areas of design, realization and testing of microsystems, sensor systems and microelectronic elements and circuits. In educational process, the services of EUROPRACTICE are utilized, especially technological realization, software and participation in specialized courses.

2 Microsystem design

At the Department of Microelectronics of the Czech Technical University in Prague, there has been pursued the development in the area of sensors and sensor systems for many years. Projects in this area are being solved. For designing microsystems, the departmental CAD facility is used [1]. The students acquire knowledge and skills in the area of computer-aided design using CAD tools, that correspond to the world standard. They use them practically for their projects. The facility offers good conditions for research activities of PhD students and research fellows as well. The design of microsystems represents very broad and complex activity. It involves design of electronic circuits or systems at all hierarchical levels, from behavioral system models to layout design. It is possible to realize all simulations on the level of electronic elements, circuits, functional structures and technological process in the facility as well. All these computer-based activities are possible thanks to professional software tools running on powerful workstations. The students get familiar with software tools like CADENCE, SYNOPSIS, HSPICE, XILINX, TCAD, ATHENA, ATLAS, UTMOS, SMART SPICE. The CAD laboratory offers many various opportunities for education of

T.J. Mouthaan and C. Salm (eds.), Microelectronics Education, 61-64.

advanced undergraduate and PhD students in courses on Microelectronics, Microsystems, ASIC Design, TCAD for Electronics, Sensor Systems, Analogue and Mixed IC Design, Exercises on IC Design. Since the CAD equipment is unique and hardly accessible for small and medium companies, dealing with electronics, this laboratory can serve as a training centre for engineers from the industry, as well. The CAD laboratory with its tools has many advantages from the point of view of education of students. It enables them to become familiar with advanced methods and techniques of IC and microsystem design, with wide-spread standard CAD tools and to continue with realization of ICs and microsystems using top European technology. Knowledge and skills acquired by the students represent good starting point for their career in advanced foreign electronic companies. From a more general point of view, it can be said that the CAD centre gives the students greater motivation for study in our specialization and improves conditions for their creative work and education. From the point of view of research, CAD laboratory offers possibilities for solving problems of electronics with powerful CAD tools. This lab opens also opportunity for collaboration with other advanced European labs in the frame of various European projects, or bilateral collaboration with the Czech industry.

3 Experimental realization and structure diagnostics

The most advantageous possibility proves to be the orientation to EUROPRACTICE. The Department of Microelectronics became member of EUROCHIP in 1992. Since 1995 it has been a EUROPRACTICE member. The department exploits these EUROPRACTICE services mainly for the purpose of software purchase and maintenance, further for realization of ICs designs. The staff members took part in various courses organized by EUROPRACTICE. The diagnostic laboratory is used for measurement of realized samples of structures and systems. The laboratory is equipped with measuring instruments and systems. The laboratory is utilized both for research and education of undergraduate and postgraduate students. It is specialized to measurement of semiconductor structures. However its variability enables other types of measurement as well.

4 Research direction examples

Of course, the projects have different character, depending on the application area, but they have at least one common point, namely concrete practically applicable results. Members of the project teams are teachers, researchers, PhD students and Master degree students. Recently the research and education in the area of microsystems has been initiated. In this area, the cooperation of several disciplines (like electronics, mechanics, optics, medicine, computer science, control engineering) is inevitable.
We will focus on questions concerning utilization of computers and computer-based tools in design, analysis, simulation and evaluation. The pursued aims can be roughly divided into following groups which are often overlapping: Development of sensor structures and

microsensors, circuits for processing of sensor signals, intelligent sensor systems for wireless transfer of measured information, miniature electronic, chemical sensor systems with ISFET transistors. Let us present several examples of realized projects. These projects represent results of research cooperation of teachers and students. Their outputs are then used in educational process.

One-chip resonance circuit. There has been designed a one-chip Si resonance circuit of a capacitor pressure sensor. The circuit consists of a variable capacitor and a flat coil, i. e. a passive LC resonator. The pressure change causes the capacitance change and thus the change of resonance frequency. The circuit as a passive resonance circuit is designed for contactless pressure measurement and wireless information transmission in medicine and other applications. The circuit has been realized as a monolithic integrated sensor on silicon. The dimension is approximately 3x3 mm. Non-standard CMOS technological steps have been examined during the sensor manufacturing. Individual modifications differ in capacitor structure, coil layout, interconnection structure, hermetization of the reference cavity, hermetization way, aluminium thickness layers.

Flowmeter with fluidic oscillator. This educational and research project has solved the design, construction, development and testing of a flowmeter with fluidic oscillator for measurement of liquid flow. This is a typical MEMS structure. A temperature resistance sensor is used for frequency indication. Liquid oscillation frequency is proportional to liquid flow velocity. A fluidic oscillator functioning as a flowmeter probe is completed with electronic circuits - Figure 1.

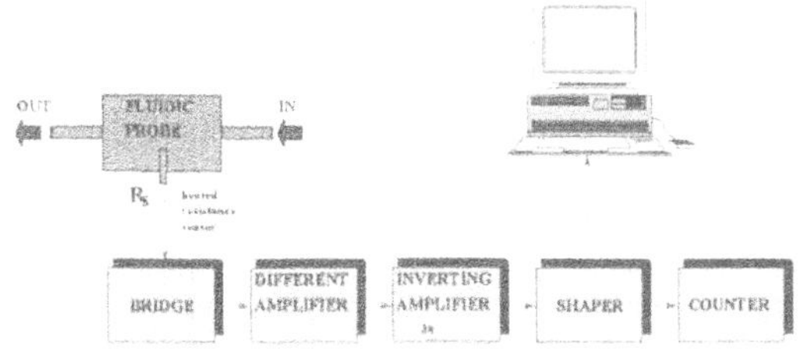

Figure 1 A Block diagram of the flowmeter

Flowmeter with dynamic calibration. The core of this project is the design, realization and parameter measurement of the flowmeter, aimed at the liquid and gas rate measurement in a pipe having small internal diameter and a variable temperature of the easured quantity. The flowmeter is controlled by a Motorola MC 68HC711 E9 single chip microprocessor. The system designed is autonomous and is equipped with the dynamic autocalibration circuits .

Temperature transducer. The project solves the design of the temperature independent phase-locked loop transducer. Two types of the CMOS temperature sensor operating in the weak and strong inversion region have been designed. The transducer is designed to convert the input signal (temperature) to the output signal (frequency) - Figure 2.

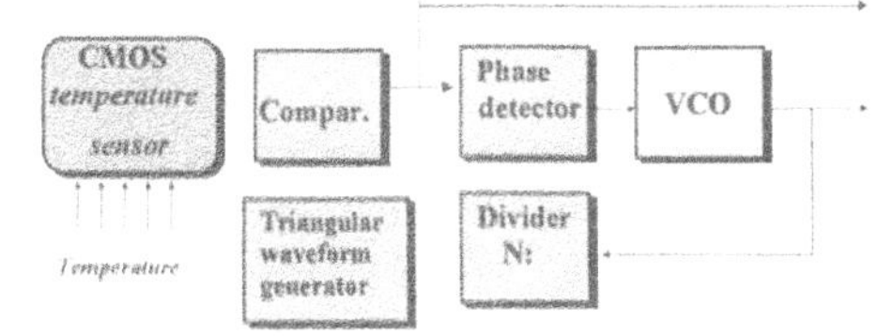

Figure 2 Temperature transducer with PLL sensor signals processing

Flow and direction monitoring sensor system. The project solves design and realization of a device for wind velocity and direction measurement. The device is designed without using of movable parts and is based on the anemometric principle. Behaviour of the digital part and the calculations are performed by a microcomputer. The measured data of wind flow velocity and direction are visualised digitally using displays - Figure 3.

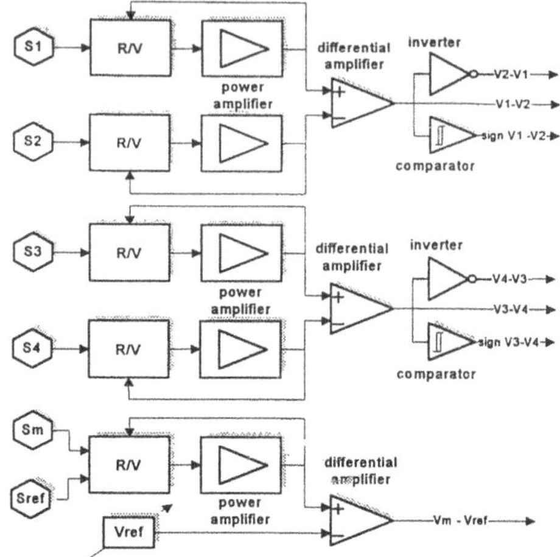

Figure 3 Flow and direction monitoring sensor system

5 Conclusion

We have presented several areas being in the focus of education and research at the Department of Microelectronics. Major part of student's education is concentrated into design. The CAD laboratory equipped with SUN workstations and world standard software supports this education. The diagnostic laboratory serves for measurement and diagnostics. The EUROPRACTICE service enables technological realizations. European programs TEMPUS and COPERNICUS support further educational development in this area. Finally, we would like to point out several factors that may influence the content not only of education in the area of microsystems, but of engineering education as a whole: *A grounding in basic* mathematics, physics and chemistry is obviously essential. However, it is necessary to distinguish between engineering and scientific (theoretical) approach. *Team skills* must be practised at every opportunity. Students must acquire *basic knowledge* of politics, sociology, economics, psychology, and perhaps most importantly, written and oral communication skills. Greater emphasis must be laid on *systems engineering*, reprezenting the integration of diverse engineering elements.

6 Acknowledgment

Research and education described in this paper has been supported by the grants No.0211/1996, No.0884/1997 and No.01237/1998 of the University Development Foundation and No.VS97046 of the Ministry of Education of the Czech Republic

7 References

[1] I. ADAMCÍK, The CAD Centre for Electronics at the CTU in Prague, Bulletin of the Faculty of Electrical Engineering, No.2., Dec. 1996 (in Czech)

TEACHING RELIABILITY IN MICROELECTRONICS

FAUSTO FANTINI*, LAURA CATTANI
University of Parma, DII
V.le delle Scienze, 43100 Parma, Italy
*also with *University of Modena, DSI*
Via Campi 213/B, 41100 Modena, Italy

BRUNO RICCO'
University of Bologna, DEIS
V.le Risorgimento 2, 40136 Bologna, Italy

MARCANTONIO CATELANI
University of Firenze, DIE
Via S. Marta 3, 50139 Firenze, Italy

1 Abstract

In this paper three different experiences of teaching quality and reliability in electronics are presented.

2 Introduction

The high reliability of integrated circuits is recognised as one of the main factors in the success of microelectronics. A large body of studies contributed in the years to this result, in such diversified fields as device physics (nowadays reliability and performance are the main concern for the development of advanced technologies), characterization methods and techniques (including electrical testing and advanced electron microscopy), experimental tests and mathematical and statistical approaches (for prediction based on accelerated tests). The efforts dedicated to the subject had significant impact on the scientific community, as demonstrated by Conferences like the "International Reliability Physics Symposium" (IRPS) organized since the early sixties by the IEEE in the USA, the younger "European Symposium on the Reliability of Electron devices, Failure physics and analysis" (ESREF) in Europe and by specific journals like "Microelectronics and Reliability". The results of the work on reliability have become essential parts of all the development programs of devices, boards and systems, but have also produced a unified new discipline deserving careful attention in

T.J. Mouthaan and C. Salm (eds.), Microelectronics Education, 65-68.

its own rights, that covers a wide area reaching with continuity that of the Total Quality Management (TQM) [1]. In spite of its obvious relevance in the real world, however, reliability is rarely included in the curricula of engineering students (and less of all in that of physicists) in universities around the world. In Italy, Reliability (and Testing) of electronics circuits and components have been inserted in the (restricted) list of basic subjects of interest for university studies in Electronic Engineering. In spite of this, however, only a few courses on the subject have actually been activated, although the pioneering experience carried out at the Universities of Parma and Bologna, and later on at the University of Florence, demonstrated the interest of both students and industry.

3 The courses

These courses are given for the five years Laurea degree in Electronic Engineering.

3.1 THE COURSE IN PARMA

The topics included in this course cover all aspects of reliability, from basic concepts of quality to reliability prediction and failure mechanisms. In particularly the course includes about 60 hours of front lessons, three practical laboratories and at least ten seminars from experts coming from the industry. The lessons cover the following topics:

- the modern aspects of quality with a detailed description of the international standards ISO 9000 [2,3] and the principles of the TQM [1];
- the statistical process control (SPC) [4], including exercices for applying the seven instruments for data analysis [5];
- the statistical bases for the reliability: probability distributions and data analysis;
- the reliability of electronic components and systems;
- the physics of failure mechanisms and the failure analysis [6].

To follow the course the students are required to have passed all the exams of electronic and microelectronic. An average number of 20 students per year followed this course since 1993. The majority of these students chose a topic either on quality or reliability for their Laurea thesis.

3.1.1 *The practical experience*

The first laboratory is done to become familiar with the curvetracer: the students are asked to characterize discrete semiconductor devices and simple integrated circuits. In the second part of this laboratory the students perform functional measurements of digital ICs by means of a programmable automatic tester.

This capability is then employed to performe practical experience of failure analysis with the purpose of training the students to face the real problems encountered in the application of microelectronics. The failure analysis includes the electrical characterization, the opening of the package and the examination by means of optical microscopy and Scanning Electron Microscopy (SEM). The samples for failure analysis come from industries or may be supplied by the students themselves. Two examples of SEM immages of failed devices taken by a group of students are reported in Figure 1. Both devices showed overstress phenomena.

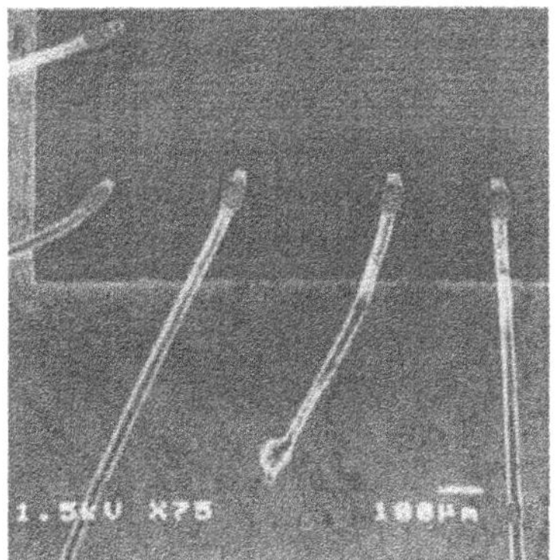

Figure 1: Examples of failure analysis performed by the students

The last laboratory is an experience of reliability prediction for an electronic board or for a simple system done by means of the RDF 93 (Handbook of Reliability Data for Electronic Components) [7]. Also in this case it's responsability of the students to choose a real circuit for the reliability analysis. A short version of the course with larger emphasis on the industrial aspects of quality is given also for the 3 years Diploma degree.

3.2 THE COURSE IN BOLOGNA

The course, held at the University of Bologna since 1992, is at the fifth (and last) year of the studies for Electronics Engineering; it's not compulsory and is chosen by about 25 students each year. It's composed of two main parts, testing of integrated circuits and reliability, in both cases treated with special emphasis on subjects related to the research activity carried out in the microelectronics Laboratories of the University. The program of the first part includes a theoretical part on the basics concepts of Computer Aided Testing (CAT), namely Fault Modeling, Automatic Test Patterns Generation, Fault Simulation, Design For Testability, Built-In-Self-Test, Self-Checking, Fault-Tolerant Architectures and Automatic Test Equipments (ATE). These lectures (about 20 hours) are completed by 15 hours of demonstrations on workstations where groups of students take an IC, generate the pattern for its testing, analyse the fault coverage and look for architectural solutions able to improve it. A brief demonstration of testing with a real ATE is also part of the laboratory work. The second part of the course starts with 6 hours on the basic concepts of reliability (statistics, accelerated tests, reliability predictions, ...) and continues with about 24 hours of lectures on reliability physics, with the aim to derive and justify the basic models used to predict IC operating life.

The physical mechanisms treated in this part may vary slightly from one year to the other, but in any case include all most important ones (hot carrier, oxide degradation and breakdown, ESD, electromigration, CMOS latch-up). These lectures are complemented by about 15 hours of laboratory work, where the students are introduced to some of the experimental techniques use to characterize reliability-related effects in electron devices (I-V and capacitance measurements, accelerated stress, deductive fault analysis,...). These demonstrations include also some work with an Eletron-Beam Tester available in the laboratory. The program briefly outlined above is normally accompanied by a number of seminars (4 - 8 hours) held by invited guests from the industry and dedicated to some of the main concepts and experiences in the field of TQM.

3.3 THE COURSE IN FIRENZE

The course, at the 5th year, is for the School in Computer Engineering, Electronic Engineering and Telecommunication Engineering. It includes front lessons, laboratory work and seminars where particular aspects concerning the industry are presented by external experts. The main topics covered by this course regard the basic concepts and methods of quality and reliability assurance, reliability of electronic components and systems, reliability prediction, methods and techniques of statistical quality control. The final part concerns the International Standards ISO 9000 and product certification. Laboratory activities in reliability prediction for electronic equipments and on the use of software tools for SPC are provided.

4 Conclusion

Although a few differences exist among the three courses their purpose is common. Three main aspects are covered:

- industrial quality, also by an extended use of experts from the industry;
- reliability and study of the failure mechanisms of the electron devices;
- testing of integrated circuits.

Based on the course in Parma a network among 15 Universities was established in the Erasmus programme, which produced the exchange of about 10 students per year since 1992 to perform their thesis abroad on reliability topics.

5 References

[1] A.V. Feigenbaum, *Total Quality Control*, McGraw-Hill, Singapore, 1991
[2] ISO 9001, 1994
[3] ISO 9004-1, 1994
[4] A. Mitra, *Fundamentals of Quality Control and Improvement*, Macmillan Publishing Company, New York, 1993
[5] K. Ishikawa, Guide to *Quality Control*, Asian Productivity Organization, Hong Kong, 1983
[6] E. Pollino (ed.), *Microelectronic Reliability*, Vol. II, Artech House, Norwood, 1989
[7] *Handbook of Reliability Data for Electronic Components* RDF 93, France Telecom, CNET, 1993

ALTERATIONS IN THE BASIC COURSE "SOLID STATE PHYSICS" IN CONNECTION WITH MICROSYSTEMS STUDY

Y. M.POPLAVKO
Microelectronics Department of National Technical University of Ukraine
37 Peremogi Ave, 252056 Kiev, UKRAINE

1 Abstract

In Technical Universities, the field of *Microelectronics* is a multidisciplinary area where "Solid State Physics" is one of the basic courses. Previously attending to microelectronics, this foundation course was in reality the *electrophysics of solids*: students studied, predominantly, electrical properties of semiconductors and metals as well as received some initial learning in dielectrics. Stimulating by industry to achieve greater functionality, microelectronics gave rise to *Microsystems*. Present and future sophisticated electronic systems will require engineers with higher degree of fundamental knowledge. Recent continual and rapid increase of Microsystems need to be reflected in teaching disciplines, and in the "Solid State Physics" especially.

2 Physical basis of Microsystems

Microsystem is miniaturized system containing sensors and actuators joined by the information evaluation part. As a result, microsystem delivers a new function due to the miniaturization that allows to get new insight on main branches of application and scientific research. Further reductions in size, which result in higher operation speed, more functionality, and lower cost are occurring at the present time.
Microsystem technology is a consistent development of micromachining which comprises the application of different technologies to form a complete microsystem. The last uses not only microelectronic devices but new types of mechanical structures which in their size range are below the traditional fine mechanics. New technique of their processing has the same advantages as microelectronics gives to electronics: microminiaturization and batch fabrication.
Microsystems started from the quartz tuning forks for frequency control in watches. Next important step was micromachined silicon structures that have the world market in pressure, temperature, gas fluent and other microelectronic sensoring. Nowadays in the Microsystems it is expected the application of new sensor materials with lower noise factor: semi-insulating III-V crystals of GaAs type, II-VI crystals of ZnO type, ferroelectric films and polar polymers.

T.J. Mouthaan and C. Salm (eds.), Microelectronics Education, 69-72.

As a typical example, the possibilities of GaAs micromachining [1] by comparison with silicon are demonstrated in the ***Appendix***. It is well known that nowadays silicon (Si) and quartz (SiO_2) are the most common crystals in microelectronic sensoring because the first one is a good semiconductor and the other is one of the best piezoelectrics. From the Tables 1 and 2, it is easy to see the GaAs type crystals combine the advantages of both silicon and quartz crystals, that is why GaAs has a large unexploded potential, especially in micromachining.
The employment of piezoelectric effect is a good chance to decrease the noise factor of sensors because GaAs sensor element uses *the change of polarization* in dielectric instead of piezoresistivity utilization *in semiconductor,* as in the case of silicon. Under anisotropy boundary conditions, in the GaAs type crystals were discovered artificial pyroelectric effect [2] and artificial volumetric piezoeffect [3]. It is clear that nowadays most of engineers need a basic understanding not only electrical but *mechanical* and *thermal* properties of solids. Physical nature of these properties are very important for correct choose of electronic materials. Moreover, some of Microsystems, such as Scanning Probe Microscopy, MicroRaman method etc., need more fundamental knowledge in atomic structure and symmetry of solids, as well as in lattice dynamics. Correspondent chapters of "Solid State Physics" should be strengthened.

3 Physics of "smart materials"

Owing to the fact that data evaluation process already gain from developments of microelectronics, the key elements of Microsystems are *sensors* and *actuators*. They used various physical effects in solid matter (semiconductors, dielectrics, metals).
Semiconductor sensors of temperature (T) or pressure (p) are based on electrical conductivity (σ) change: scalar variables δp or δT are indicated through the effects of piezoresistivity $\delta\sigma(\delta p)$ or far-infrared photoconductivity $\delta\sigma(\delta T)$. However, conductivity is accompanied with noise.
Dielectric sensors have lower noise factor because they use the change of spontaneous polarization (P_s). The P_s variations provides pyroelectric effect ($\delta P_i = \gamma_i \delta T$) as well as volumetric piezoeffect ($\delta P_i = e_i \delta p$), both of them are vectorial variables. Therefore in students education the emphasis should be made on the nature of dielectric polarization in the non-central dielectrics, especially to their possibilities to transform thermal, mechanical, optic and other influences into electric signals.
The most of interest for Microsystems are materials that have ability to perform both sensing and actuating functions. Piezoelectrics, for instance, are capable of acting in this way: such materials utilize feedback loop enabling them to both recognize the change and initiate an appropriate response through the actuator circuit.
In the USA materials with such capabilities are named as *smart materials*, Russians named them as *active dielectrics*. That is why a new version of "Solid State Physics" should pay more attention to the physical background of these dielectrics.
In the Report a new version and content of "Solid State Physics" will be proposed with the attempt to consult nowadays and future development of Microsystems and Microelectronics. In sensoring, by way of example, the variations of some internal physical parameters of solids are usually used.

Some dielectrics used in *Microsystems* can transferee mechanical and thermal properties into electrical once (sensoring) or vice versa (actuators). For example, piezoelectricity binds two main properties of solids: electrical and mechanical ones. In the same way, pyroelectricity is based on thermal-to-electrical signal transverse, etc. Therefore, along with the main electrical properties some *elastic* and *thermal* properties of solids should be previously examined. As to the *optical* properties of dielectrics, they reflect in the most responses the possibilities of field controlling of crystals parameters.

4 Integrated ferroelectric films

Recent times have seen a tremendous increase in ferroelectric and other "active" dielectric films applications for use in displays, sensors and memory devices. Integration of these films with semiconductor chips permits to combine the unique properties of ferroelectrics with the advantages of semiconductor IC chips.
Integrated ferroelectric-semiconductor devices are really a new chapter of micro-electronics: "active" dielectrics are multifunction elements which essentially increase the possibilities of microelectronic processors while semiconductor chips provide low cost, large capacity, high density, amplification and logic functions.
Outstanding progress in the integrated films applications stimulates new fundamental physical studies of such a complex system. Topical problem is to explain the change of dielectric polarization and losses mechanisms while bulk dielectric is transformed into the film integrated with other material substrate. The change in *bulk - film* properties could be either favorable or adverse factor for electronic devices. This knowledge will be used for essential improvement of film processing.
In the course of "Solid State Physics", theoretical approach to ferroelectric integrated film properties is based on 2D (two-dimensional) crystal concept with a peculiar lattice dynamics, for the ferroelectric soft lattice mode especially. Moreover, for *2D* crystal its surface layers and near-electrode interfaces have marked effects on electrical properties. Integrated film is rigidly bounded with substrate and electrodes that have a quite different elastic and thermal properties. Being sintered at high temperatures, thin film at the normal conditions should be stressed due to the difference in thermal expansion coefficients between film and its sandwich-type environment. Therefore, film could be tensioned or compressed. Stressed in plane, ferroelectric or paraelectric film could even change its symmetry but in any case changes its electrical properties.
There are some important distinctions in *dynamic conditions* causes for the bulk - film discrepancy in properties. The point is that the film bound by substrate can not fully realize its dynamic strains that would be electrically induced by piezoelectric effect or electrostriction. All these effects should be reflected in "Solid State Physics".

References

[1] H.L. Hartnagel. Present limitations and component requirements. Proc. 24-th European. Microwave Conference, 1994, pp.151-154.
[2] Y.M.Poplavko, L.P.Pereverzeva. Pyroelectricity and other effects in partially clamped crystals. *Ferroelectrics*, 1992, Vol. **130**, pp. 361-366; *Ferroelectrics*, 1994, Vol. **153,** pp. 353-358.
[3] Y.M.Poplavko, L.P.Pereverzeva. Volumetric piezoelectric effect and artificial pyroelectricity in GaAs. Proc. Europ. Symp. on GaAs, Paris, 1996, p4A2.

Appendix.

Table 1. Microsensoring with GaAs type crystals, index (*) shows the Si crystal possibilities

Sensoring parameters:

Temperature*	T
Stress *	X
Electric field*	E
Light*	Ψ
Pressure*	p

Utilized effects:

(i) based on electric charge generation, and conductivity $\sigma(T, X, E, \Psi, p)$ or resistance $\rho = 1/\sigma$ measurements	Thermoresistance* Piezoresistance* Photoresistance*
(ii) based on electric charge separation: $q(T, X, E, \Psi, p)$, and electric current dq/dt or electric potential measurements	Piezoelectric effect Artificial pyroelectric effect Artificial volume piezoeffect
(iii) based on light generation or light transparency control	Photoluminescence Piezooptical effect

Remark: When conductivity is used for sensoring, it means in semiconductor a non-equilibrium electrical *charges generation.* Any conductivity process produces some noise.
If electric polarization is used for non-equilibrium *charge separation,* the sensor is dielectric (or it is very close to dielectric), and noise factor should be lower.

Table 2. GaAs type crystals as materials for microsensoring in comparison with silicon

GaAs peculiarities:

low sound velocity	(0.5 /Si)
high density	(2.2 /Si)
high thermal resistivity	(3.1 /Si)
high thermal expansion	(2.4 /Si)
strong etching anisotropy	

GaAs advantages:

possibility to use in optoelectronics
piezoelectric effect
artificially arranged pyroelectric effect
artificial sensoring of pressure
high working temperatures
low microwave absorption
low thermal conductivity
possibility to use „stop etching" AlGaAs layer

GaAs disadvantages:

high cost
difficulties in oxidation (to form protection layer)
difficult bolding
fatigue from gradual diffusion processes
lower chemical endurance
lower mechanical endurance (fragile, brittle):

hardness	0.7 /Si
fracture strength	0.5 /Si
toughness	0.5 /Si
less wear rate	

Remark: In the AlGaAs alloys mechanical properties are much better than in GaAs.

USE OF SYSTEM APPROACH IN MICRO/OPTOELECTRONIC DEVICES DESIGN

P. VIGIER, C. BERTHELEMOT-AUPETIT and J.M. DUMAS
Centre National de Formation en Microélectronique (CNFM)
Pôle Limousin de Microélectronique (PLM)
Ecole Nationale Supérieure d'Ingénieurs de Limoges (ENSIL)
Parc d'ESTER Technopole
87068 LIMOGES Cedex, FRANCE

1. Introduction

In order to design high bit rate optical communication systems, engineers need very sophisticated and expensive components, especially in optoelectronic emitter and receiver blocks [1]. A usual budget study, based on Gain, Bandwith and Signal Noise Ratio, leads to specifications too much drastic and consequently to extracosts. The system approach is the way to determine if a device will be efficient for the system functions and therefore allows to define more realistic specifications. In this approach, the characteristic measurements are eye diagram and bit error rate.
A first step to insert this point of view in the ENSIL educational program was the definition of three student projects.

2. Gate and Drain Lag effects

When investigating parasitic effects on III-V FET (HEMT)-based IC, gate and drain lag phenomena appear to be among the main ones. They are a good examples in order to initiate students about using different approaches for a problem.

2.1 PHYSICAL PHENOMENON

Materials used for microelectronics devices are not physically perfect. There are trap mechanisms disturbing the current during transient operation. For example, the establishment of a current may be majored or reduced by an additional current due to the charges which may be trapped or released. As a result, a digital signal will be degraded when processed by a III-IV device.

T.J. Mouthaan and C. Salm (eds.), Microelectronics Education, 73-76.

2.2 ELECTRICAL BEHAVIOUR

The study of the electrical behaviour is the simplest way to detect parasitic effects. Two different approaches are possible, one in the time domain and the other in the frequency domain.

2.2.1 Time domain

In order to evaluate the impact of these effects, the most evident idea is to measure the degradation of a pulse representing a bit (duration function of the bit rate). The different load impedance possibilities are taken into account through two different drain current measurements.

At first, a voltage pulse is applied to the drain with a constant voltage applied to the gate. This is the drain lag measurement. It corresponds to an infinite load impedance and degrades the transistor gd. (see Figure 1 left)

Second, a voltage pulse is applied to the gate with a constant voltage applied to the drain. This is the gate lag measurement corresponding to a near-zero load impedance and degrades the transistor gm. (see Figure 1 right)

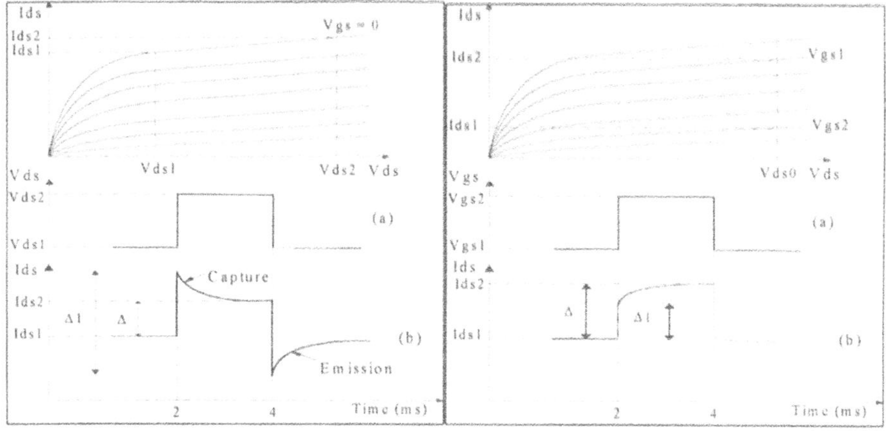

Figure 1 : Gate and drain lag time domain electrical behaviours

2.2.2 Frequency domain

Those effects are translated in the frequency domain by the frequency dependence of intrinsic characteristics of FETs (gm and gd). Each one has a value at low frequency and another one at high frequency.

2.3 EXPECTED SYSTEM IMPACT

For bit duration overlapping the trap time constants, we expect to observe an eye diagram penalization. on the other and, when the bit duration is lower than the trap time constant, we do not expect a strong impact.

3. System simulation approach

In this project, two students developed models to simulate the gate lag of a transistor. As it can be seen in Figure 1, the gate lag effect may be interpreted as the sum of the normal response of the transistor and the response of a low pass RC filter. The drain lag effect is the sum of the normal response and a high pass filter response. For this reasons, a first step was the simulation of low pass and high pass filter using AMI code. It was useful to validate both simulation results and measurement procedures. In Figure 2 a good correspondence is observed excepted a little deformation due to the data generator at high bit rate.

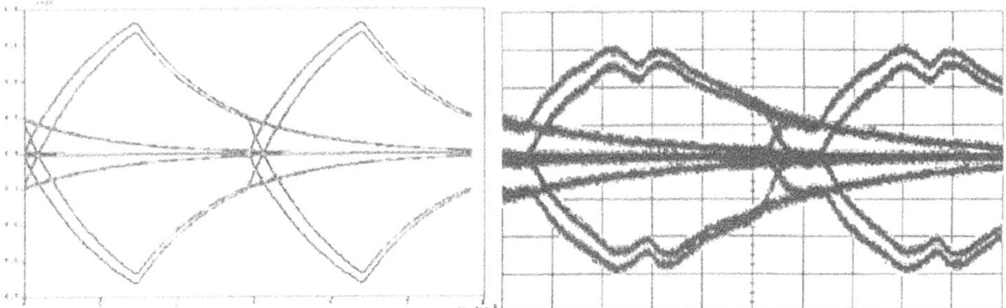

Figure 2 : Simulated and measured responses of a low pass filter at 34 Mbit/s

The second step consisted in the definition of different models to simulate the gate and the drain lag. In a first approach, those phenomena are linear, so they are correctly described by a simple transfer function (see (1)).

$$H(s) = \frac{\left(1 + RCs\frac{1+k}{k}\right)}{1 + RCs} \qquad (1)$$

Figure 3 shows the system impact for gate lag when the bit duration is equal to two or three times the constant time (left) and the impact when time constant is several orders of magnitude greater than the bit duration (right). The simulation results look as expected for low bit rate while a level shift is observed at high bit rate.

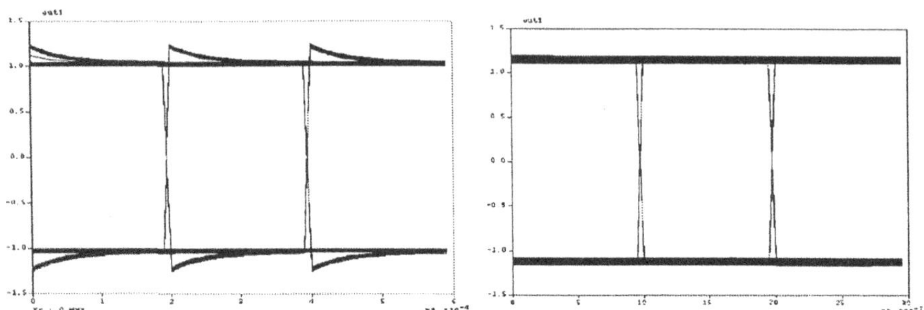

Figure 3 : System impact for gate lag at low and high bit rate

4. Sub system measurement

A measurement setup has been developed by students in order to validate the simulation results for both AMI and NRZ codes. The measurements of low pass and high pass filters pointed out the necessary bandwith needed to transmit a digital signal.
Today, this measurements are not yet optimised for the NRZ code because bias tees. This point is under solving.

5. Conclusion

We made the link between high speed devices and optical digital systems. This has been done by means of experimental together with simulation exercises carried out by students. Their project are now under operation and have been introduced in the teaching curses, thus illustrating the basic lectures.

6. References

[1] J.M. Dumas, P. Vigier, C. Berthelemot and J.P. Cances, « Introduction of semiconductor devices into communication systems : teaching examples », Proc. of the European Workshop Microelectronics Education, World scientific 1996, pp 253-256

TEACHING DEVICES AND TECHNOLOGY WITH SIMULATION ON PC AS A DIDACTICAL APPROACH.

Two dimensional device and process simulation with MicroTec in a teaching environment.

A.J.Mouthaan*, M. Obrecht
*Twente University of Technology, Enschede the Netherlands,
a.j.mouthaan@el.utwente.nl
University of Waterloo, Canada,
Obrecht@vlsi.uwaterloo.ca

1. Abstract

Semiconductor devices and IC technology are considered 'difficult' subjects in a curriculum of Electrical Engineering. The reasons are the strong focus on physics and to some extent mathematics. Engineers by nature are not biased towards theory, and the avalanche of formulas they are confronted with somehow obscures their real understanding of the subject and the true 'engineering' nature of semiconductor devices. The use of process and device simulation brings engineering aspects in the course: students do experiments and are not solving formulas. This way the didactics better suit the training of engineers. This paper gives an example of teaching through simulation using the PC process and Device simulator MicroTec from Siborg.

2. Introduction

This simulation based introduction to semiconductor devices tries to teach undergraduates in electrical engineering the fundamentals of our common semiconductor components by actively using simulation. Active means that students do simulations to get to grips with device physics. There are two main reasons for this concept.

• Engineers differ from pure scientists in that they have a desire to create tangible 'things'. By nature they are not attracted to the very strong physics oriented approach and the mathematics needed to understand and create semiconductor devices and integrated circuit processes. Semiconductor device courses are popular by too few engineering students. The link to real 'things' can be provided by simulation; electrical characteristics can be derived from it with relative ease, covering many different biasing modes and applications can be made in electronics through a coupling with circuit simulators.

• In device and process research simulation has become the tool that engineers and

T.J. Mouthaan and C. Salm (eds.), Microelectronics Education, 77-80.

scientists use. Simulation supported by a good understanding of devices is the prime analysis method. Simulation can be extended to support the understanding of basic phenomena. In this way students learn to use the method and learn the fundamentals of our devices.

There is a large range of very good books on the fundamentals of semiconductor devices. Still it is common experience of teaching staff at universities that many students are put off by the subject because of the mathematics and physics involved. The mathematics and physics are not difficult but they are overwhelming: there is no book with less than 300 formulas. That most of the formulas come from less than five basic relations is not seen by most students unfortunately. Yet semiconductors are so much fun and are a true engineering subject also: engineers should not leave it to the physicists and chemists.

This approach tries to increase the engineering challenge in our subject. Modern PC's are strong enough to do our simulations and several high quality simulators are now available on the market for PC's. Simulation is the logical sequel to the slide rule and the calculator as the engineering tool. In this project students are encouraged to do simulation exercises and some larger projects are described that can function as examination assignments.

3. Teaching semiconductor devices as compulsory subject to undergraduates; aims and methods.

For courses selected for the undergraduate programme the objectives should be clearly defined within the context of this core programme; they should not simply be "Introduction to...". The core programme should teach methodologies, fundamental concepts, technological concepts in order to make all graduates, irrespective of their further specialisation, qualified to work in our profession as independent engineers able of judgement and capable of inventive and integrative engineering practice.

The objectives of a course on devices are derived from its context (support the education in design and methods for microelectronic design) and it should present fundamental concepts and technological principles that are of lasting value to all our students. Below the objectives for this course are summarised by stating the capabilities that students, after completion of this course, are expected to poses.

- students must understand the use of modelling of components, describing the behaviour of the components in terms that are relevant to the actual design environment;

- students must be able to present the first order models of (homo-) junction diodes, (homo-) bipolar transistors and MOS field effect transistors in the context of the dc, large signal and small signal (low and high frequencies) design level;

- students must be able to relate (qualitatively) the components of these models to the physical and geometrical properties of the devices;

- students must appreciate the major challenges of device development in the context of analog, digital and microsystem design.

Key element in modelling in whatever way is to indicate what the limit of the applicability of the models is. Every model has a limit and the number of students that believe that simulation is reality is alarming. The solution is not to limit simulation, but to teach it. For device research, simulation is the only viable tool at universities and luckily through industrial contacts or on their own behalf many can do clean room verification and experiments.

In this course simulation is the exercise method. A two dimensional process and device simulator MicroTec, marketed by Siborg in Canada, is part of the text and all examples in the text are simulated with this simulator.

4. Examples of simulations that enhance understanding.

First introduction to technology:

1. Implant Boron with an energy of 100 keV and a dose of 1e12 cm-2. Determine the peak concentration and relate that to a Gauss approximation for the implanted profile. Out diffuse at 100 C for 1 hour. Approximate the characteristic length for diffusion and compare it to the simulation results.
Simulation gives a feeling of the speed of diffusion, parameters of characteristic length and gives a feeling of numbers for implants.

2. Use the device simulator to determine the carrier concentrations and Fermi level in the example above for zero bias. Also determine the potential in the structure. Interpret the result of the potential. Explain why the current through the structure is zero while we have an electric field.
Simulation makes the relation between doping and Fermi level and in this case also inhomogeneous doping and resulting electric field. The balance between drift and diffusion is illustrated.

3. Using an example file, change the base doping in the neutral base region of a pn diode with an extra boron implant such that the electron current density is 100 times larger than the hole current density.
This teaches the students that the currents are minority currents that are determined by the minority concentrations.

4. Investigate the influence of the recombination velocity at the metal contacts on the current densities. Increase one of the doping levels to such a level that the influence of the contact on that side is minimal.
This teaches the students that the diode currents are diffusion currents that are determined by the diffusion length and the recombination velocity/time.

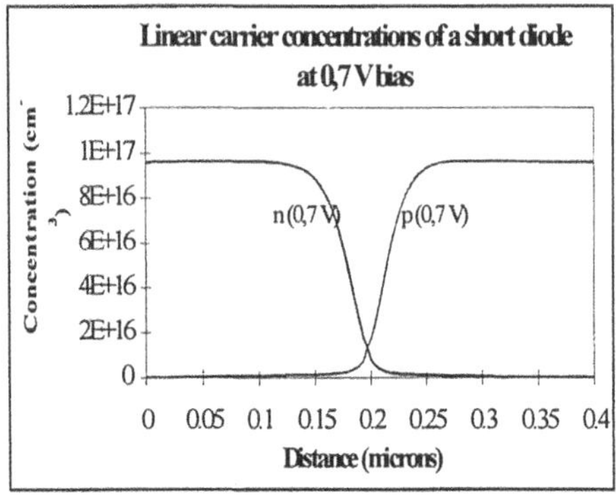

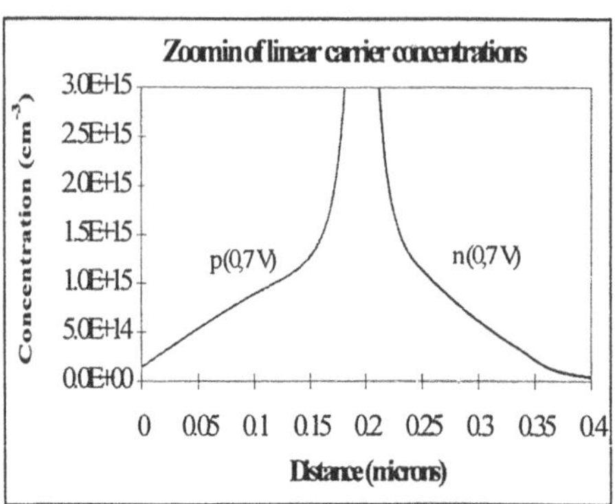

5. Adjust the base doping of a bipolar transistor (dose and depth) and the concentration of the epi-layer in such a way that high injection effects as well as β improve (use the project file).
Here students learn the trade offs and the physics behind high injection and the value of β.

6. Derive the model parameters of the MOS transistor with homogeneous bulk doping and L=0,5 μm and L=1 μm, for the saturation, linear and subthreshold region.
This teaches students how to derive model parameters and shows the effect of a short channel.

7. Study the transition from weak inversion to strong inversion for the example transistor (V_{DS}=0,1 V) by studying the gradient of the electron concentration in the channel.
Here students see the drift and diffusion behaviour of the current in both regimes.

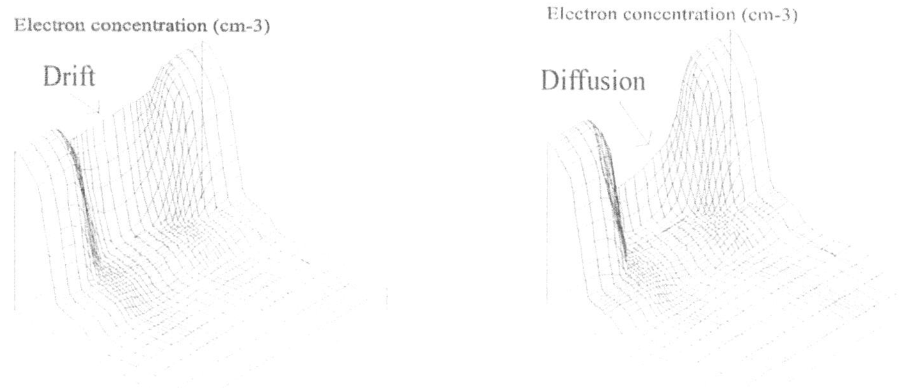

5. Reference

[1] Microtec Users manual, Siborg.

Session B

EMERGING FIELDS

A MICROENGINEERING CURRICULUM AT EPFL

Philippe Renaud
Swiss Federal Institute of Technology, Lausanne
DMT-IMS, EPFL, CH-1015 Lausanne, Switzerland

1 Introduction

The Swiss Federal Institute of Technology at Lausanne (EPFL) provides an full university level education curriculum in microengineering since 1978. At the beginning, the teaching was provided by professors attached to various departments (faculties) of EPFL such as mechanical engineering, physics and electrical engineering. In 1989, the department of microengineering was officially created at EPFL; this is the only university faculty in Switzerland which is exclusively oriented to education and research in microengineering. However, there are other institutions which are also active in microengineering and which provide punctual undergraduate and graduate courses in microengineering. In particular, the University of Neuchâtel is sharing part of the education program with EPFL and also delivers doctorates in Microengineering.

According to our definition, the microengineering is « the art of creating, manufacturing or using miniaturized components, devices or systems that are produced in series ». The primary characteristics of this domain is its multi-disciplinary nature, since microtechnology systems implies mechanical, electrical, informatics, optical elements and various materials. Therefore, in microengineering, the manufacturing aspects of a system are considered as important as its final function. Examples of microtechnology products are: sensors and measuring instruments, medical instrumentation, optical instruments, small motors, watches, small and micro-robots, portable telecommunications, computer peripherals, consumer products, toys, etc.

T.J. Mouthaan and C. Salm (eds.), Microelectronics Education, 83-85.

1st cycle (year 1 and 2)	Non- specific Specific	mathematics, physics, chemistry, programming, optics, material science electrotechnics, circuit modeling, electronics, mechanical design, CAD
2nd cycle (year 3 and 4)	General Projects Integrated products Production techn. Applied photonics	product design, technology management production and assembling techniques, electronics and microelectronics, sensors and microsystems, microcontrolers, mechatronics signals and systems, automation two semester projects micro-electronics, microsystem technology option courses robotics and micro-robotics, assembly techniques option courses optics, lasers, options courses
Diploma work	A 4 month project which is performed either at: - - - - -	 a research group of the department the University of Neuchâtel an other department of EPFL industry abroad (exchange student)
3rd cycle (doctorate / post-formation)	Admission Doctorate school Post-formation	admission test specialized courses scientific seminars short course seminars

Table 1: Microengineering studies at EPFL

2 The study plan

The challenge for an academic curriculum in microengineering is to reach a high level in several disciplines while having broad competencies oriented to microtechnology products. The EPFL way (see table 1) is to give a strong emphasis to a multidisciplinary education consisting of courses from various relevant disciplines to microengineering and, in parallel, to involve the students in several projects that are managed in a rather independent manner. In the forth year, the student have a choice between three orientations: integrated products, production technology and applied photonics. Although these are not deep specialization, these orientations correspond to different approaches of the field. The four years studies are completed by a diploma work of four months duration. It is usually done within one of the research group of EPFL or University of Neuchâtel, but can also be done outside of the Department at EPFL, in industry or abroad (under the supervision of one EPFL professor).

About 20% of the student will continue some education after their diploma, either as research assistants or doctorate students. A doctorate title in « microengineering » is delivered by EPFL.

3 Microsystem education at EPFL

Since about 1993, the microsystem related activity has rapidly grown at EPFL: (1) a new research institute the « Institute of Microsystem » has been created, (2) new clean room facilities with processing equipment for microelectronics and microsystems is being installed. This has somewhat influenced the educational capabilities and also triggered the interest of students for microsystem technology.

The policy is to let a direct access of the undergraduate students (2nd cycle projects and diploma) to the processing equipment. An introductory practical session is given to those who have chosen the « integrated products » orientation. In this session, every student has to manufacture by himself a silicon thermal actuator, to model it and to characterize it. Then, for specific projects, they are directly supervised and trained by some senior researcher or Ph.D. students. This way, a lot of responsibility if left to the students. Until now it has been quite successful and never lead to major problems.

MECHANICAL ENGINEERING COMPONENT IN MICROELECTRONIC SYSTEMS CURRICULUM

ANDRZEJ RUCINSKI, ROBERT JERARD, GERALD SEDOR, TAMÁS VISEGRÁDY
College of Engineering and Physical Sciences
University of New Hampshire, Durham, New Hampshire 03824, U.S.A.

1 Introduction

The need for multidisciplinary education in microelectronics systems is primarily driven by the evolution in scaling, when physical phenomena and not electrical circuit theory prevail, and the revolution in packaging, which calls for spatial, electrical, and temporal considerations. However, it remains difficult to incorporate a multidisciplinary component in traditional electrical engineering curricula despite a positive feedback from both funding and accreditation agencies as well as industry in the United States [5]. The latter, particularly in the area of CAD, evolves from a single technology solution towards system engineering, and beyond [3]. The paper describes a collaborative engineering approach being implemented at the University of New Hampshire which is based on two premises: (1) the emulation of a real industrial environment, and (2) challenging multidisciplinary projects as a driving vehicle [2].

2 Collaborative engineering approach

The nucleus of the University of New Hampshire collaborative engineering program is a lecture and seminar course taught jointly by the mechanical and electrical departments, complemented by industry seminars, and combined with several projects. Projects are accomplished by a mix of management, electrical, and mechanical engineering students. The last component is a new one in the current version of this revolutionary course. The design is pursued using a "parametric design" approach. This approach separates the design domain from fabrication domains, both in electrical and mechanical parts, and promotes design expandability and scalability. The lecture and seminar components of the course include such topics as ISO9000, the design process, project organization and management, concurrent engineering, reliability and professional ethics. Lectures are presented by faculty members as well as senior industry representatives. It is imperative in the collaborative engineering program that "real" projects, either research or industrial oriented, are promoted.

T.J. Mouthaan and C. Salm (eds.), Microelectronics Education, 87-90.

3 Case study: Stacked MCM system

3.1 PROJECT SELECTION AND OBJECTIVES

As a case study, an embedded microelectronic system for reusable space vehicles [1], [4] is selected for the stacked MCM implementation[1]. The selection is attractive for several reasons: first, the embedded system offers a sufficient level of complexity combined with the requirement for ruggedness and miniaturization. Secondly, MCM packaging necessitates in addressing thermomechanical issues. Thirdly, the topic appears to be extremely attractive for both mechanical and electrical engineering students because of its visibility and importance.

3.2 PROJECT DESIGN

Functionally, the MCM system is collecting sensing data through ADC conversion from up 32 sensors at 10KHz each. The information is collected by a system and then downloaded to two redundant central units through an FDDI network. The MCM system is designed around a PowerPC 403GC microcontroller emulated by a XILINX FPGA device. The system is in VHDL and synthesized. The physical implementation requires a stack of four MCM boards: one digital, two analog, and the power board connected with flexible cables.

3.3 MULTIDISCIPLINARY COLLABORATION

The mechanical engineering part of the project adds an additional complexity to the already ambitious design. Interestingly enough, mechanical engineering students are setting boundary conditions on heat dissipation, mass, and sizes. For example, the whole system has to weight no more than 5 pounds. In addition, design specifications for the MCM stack depend on a flight profile of the space vehicle and required reverse heat flow dissipation. Data collection phase (either in space or in upper atmosphere) assumes that the heat is generated by the device followed by the storage phase when the primary heat source is located outside the device (on ground).

The thermal requirement of the system indicates that a system for isolating electrical boards from the outside temperature is needed after the craft lands. A thermoactuator is used to move a conduction path composed of two wedges, one of aluminum and the other of brass, into contact with the aluminum shell when heat transfer to the outside environment is required and retracted when thermal isolation is desired. However, the system also needs a thermal path to dissipate up to 20 watts generated by the boards. To accommodate this requirement, a phase change material is applied using the Aerogel insulation as well as the disengagement of the thermally activated conduction

---

[1] We wish to acknowledge and thank Sanders, a Lockheed Martin Company for their support for the project and personally thank for his support and assistance Mr. Raymond Garbos, Engineering Fellow, Aerospace Division at Sanders

path. As the heat is transferred into the phase change material it absorbs all the thermal energy to change phase and thus its temperature remains virtually the same during the process, keeping the boards' temperature at an acceptable level. The high specific heat of the phase change material also acts as an energy storage device.

3.4 PROJECT ORGANIZATION

A total of 11 students participated in the project in the fall of 1997. Among them were 6 MEs, 4 EEs, and one MBA student. Although the students had different background and design perspectives, as depicted in Figures 1 and 2 respectively, the common design goal has resulted in a unified multidisciplinary team capable of delivering a working microelectronic system.

4 Conclusions

The described program coincides with the use of CAD systems in electrical engineering curricula [3]. This process started with design activities where student designs are focused on a single technology such as VLSI. The next generation of designs is implemented with several technologies, e.g. printed circuit boards populated with COTS parts. The current trend of developing complex microelectronic systems requires multidisciplinary approach.

This approach generates both technical and non-technical problems. Technical difficulties are related to challenges of a design itself, which are typically non-trivial and prototype oriented. Non-technical issues are risen by limits of time and schedule as well as inherent existing barriers among departments within an academic institution [5]. The latter causes difficulties the formation of a multidisciplinary design student team in a typical academic environment since curricula are constructed traditionally around disciplines i.e. departments. It is also concluded that the importance of engineering management and documentation system is underestimated, which calls for more aggressive involvement of faculty and students from business and management.

5 References

[1] T. K. Mattingly, "A simpler ride into space", *Scientific American*, October 1997

[2] A. Rucinski, B. Dziurla-Rucinska, and A. A. Forrest, "Curriculum in engineering team management with strong scientific and high-tech design components", *Proceedings of the 3rd East-West Congress on Engineering Education*, Gdynia, Poland, September 1996

[3] A. Rucinski and F. Hludik, "An evolution of an electrical engineering curriculum: From CAE fundamentals towards concurrent engineering", *Proceedings of the 1996 Mentor Graphics Users Group International Conference*, Portland, Oregon, October 1996

[4] R. Schroer, "Impact of avionics on the future and the future of avionics", *IEEE AES Systems Magazine*, August 1997

[5] *ABET Engineering Criteria*, http://www.abet.org/

Figure 1 Render Version of the Blueprint

Figure 2 Housing Prototype

SURFACE MICROMACHINING AND ELECTRICAL CHARACTERIZATION OF POLYSILICON MICROCANTILEVERS

A.M. IONESCU[1*] , P. MORFOULI[1], N. MATHIEU[1],
J. BRINI[1], N. GUILLEMOT[1,2] and J.-M. TERROT[2]
[1]ENSERG-INPG, 23, rue des Martyrs, BP 257, 38016, Grenoble, France
[2]CIME-INPG, 46, avenue Félix Viallet, Grenoble, France

1 Abstract

The surface micromachining and the electrical characterization of polysilicon microcantilevers are presented as practical works, conceived as an introduction to MicroElectroMechnicalSystems (MEMS), for students in electrical engineering at the National Polytechnic Institute of Grenoble, France. The aim of these practical works is to illustrate the total compatibility of sensor and actuator technology with MOS integrated circuits.

2 Introduction

Electrostatic micro-devices (micro-cantilevers or micro-bridges) are popular due to their potential various applications: microrelays, microswitches, optical mirrors with large tilt angles, nonvolatile memory elements, etc [1-3]. Their microminiaturization is related to the fabrication techniques which adapt IC-processing (namely film formation, doping, lithography and etching) [4-7]. These practical works show essentially how basic steps associated to the fabrication of a MOSFET can be adapted for the surface micromachining of polysilicon microcantilevers. The students can investigate each fabrication step via quasi-industrial physical characterization methods. Moreover, an advanced 2D technological simulator (ATHENA/Silvaco) provides a better understanding of the processing. Finally, the static electrical characterization of microcantilevers is used in order to evaluate the mechanical and electrical properties of polysilicon. Note that MOSFETs fabricated with the same technology (2μm - NMOS) are also tested by students, as a last step of these practical works.

3 Fabrication and control

The technological process used for the surface micromachining of microcantilevers is described in Fig. 1; it uses SiO_2 as a sacrificial layer and Si_3N_4 as an insulator. The main technological steps are : (i) deposition, photolithography and etching of Si_3N_4, (ii) Si doping, (ii) wet thermal oxidation, (iii) deposition, doping, photolithography and dry etching (SF_6 plasma) of polysilicon, (iv) SiO_2 wet etching (HF). Microcantilevers and microbridges with various widths, W (=2 up to 12μm), and a fixed length, L(=80 μm), are fabricated with masks commonly used for MOS transistor process. Various control

* A.M. Ionescu is now with LETI-CEA, Grenoble, France

T.J. Mouthaan and C. Salm (eds.), Microelectronics Education, 91-94.

characterization techniques are carried out during or after the fabrication: (i) *ellipsometry* - the beam depolarisation of a low power (1mW) He-Ne laser, after reflection on a SiO_2/Si structure allows oxide thickness extraction, (ii) *'alpha-step' technique* - after lithography and etching of various layers, this measurement provides the cross section profile and highlights cantilever bending (it demonstrates the 'free-standing'), Fig. 2, (iii) *4-point probe measurement* - it serves as a fast monitor of wafer and polysilicon average doping, (iv) *Scanning Electron Microscopy (SEM)* - its higher magnification, resolution and depth of field compared to optical microscopy make it the ideal tool for the final verification of the 3-D structure of the microcantilever, Fig. 2.

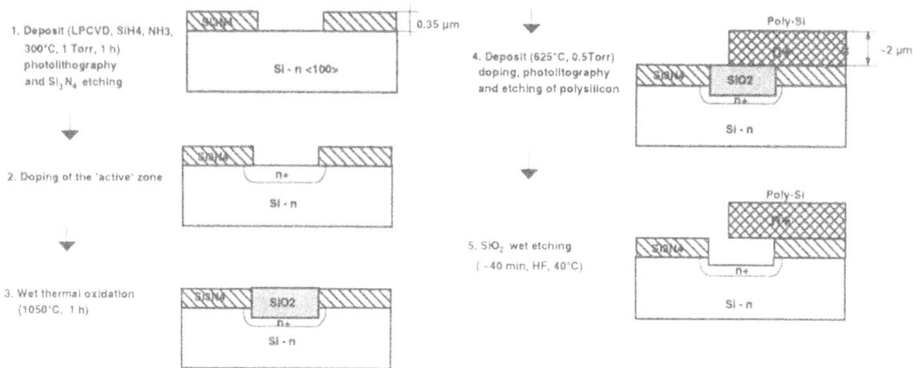

Fig. 1 Simplified technological process sequence for the surface micromachining of polysilicon microcantilever using SiO_2 as sacrificial layer and Si_3N_4 as an insulator.

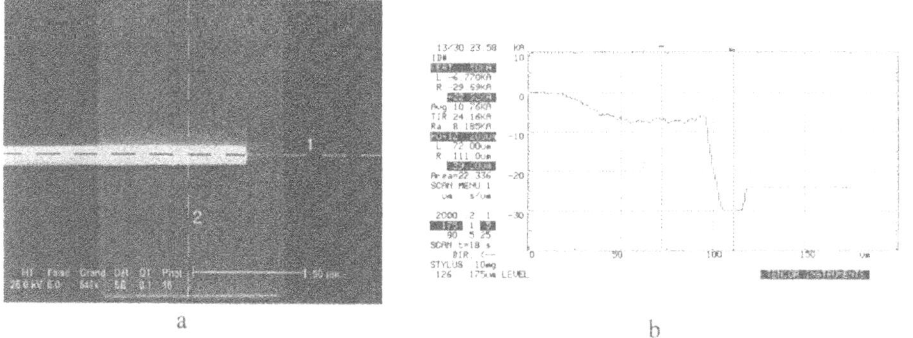

a b

Fig. 2 a. SEM image of a typical polysilicon cantilever, b. alpha-step profile along '1' axis in Fig. 2a (microcantilever bending highlighted).

4 Two-dimensional (2D) simulation of the fabrication

Some of the main technological steps are simulated in detail with ATHENA (Silvaco) simulator. A critical point is the simulation of the wet oxidation for which students can tune temperature, duration and HCl concentration (%), in order to fill exactly the sacrificial region. The well known beard beak effect, responsible in our process for supplementary mechanical stress in the Si_3N_4 layer, is also highlighted. 2D simulation stands as a powerful tool which reveals in few minutes all the fabrication sequence.

5. Electrical characterization

The deflection, y(x), of a cantilever beam due to a voltage difference, V_A, applied between the cantilever and the substrate, mirrors both the device geometry (L, W, the distance, e, between the substrate and the cantilever) and the material mechanical properties (i.e. Young modulus, E) [9]:

$$EL\frac{\partial^4 y(x)}{\partial x^4} = \frac{\varepsilon W V_A^2}{2(e - y(x))} \qquad (1)$$

Taking into account that tip deflection can be found from the balance between electrostatic and electric forces, the pull-in voltage, V_{pi} is easily calculated following the models of Tas et al.[8] or Osterberg and Gupta [9]:

$$V_{pi} = \sqrt{c\frac{Ee^3t^3}{\varepsilon L^4}} \qquad (2)$$

where t is the polysilicon thickness and coefficient c is 0.22 for cantilever beams and 11.9 for doubly clamped cantilevers (microbridges). A plot of typical deflections as functions of Young modulus and microcantilever geometry is given in Fig. 3. This calculation does not take into account the cantilever bending under its own weight. Note that for our process and structures V_{pi} is in the range 3-8 V for microcantilevers, and 20-50 V for microbridges. A HP-4155A parameter analyser is used for this static characterization which makes the experimental procedure not only highly fast and accurate but also extremely attractive for students. Typical measurements are presented in Fig. 4.

Complementary measurements of the structure capacitance as a function of voltage and frequency are expected to give a more understanding of the various applications of this kind of structures and a more accurate electrical characterization. This aspect is currently under investigation and, for the moment, not fully available for students during the practical works.

Note that n-MOSFETs fabricated using the same technology are investigated by conventional electrical characterization methods (I_D-V_G and g_m-V_G curves allowing threshold voltage, V_T, channel carrier (electron) mobility, μ_n, and transverse field mobility reduction coefficient, θ_G, extractions).

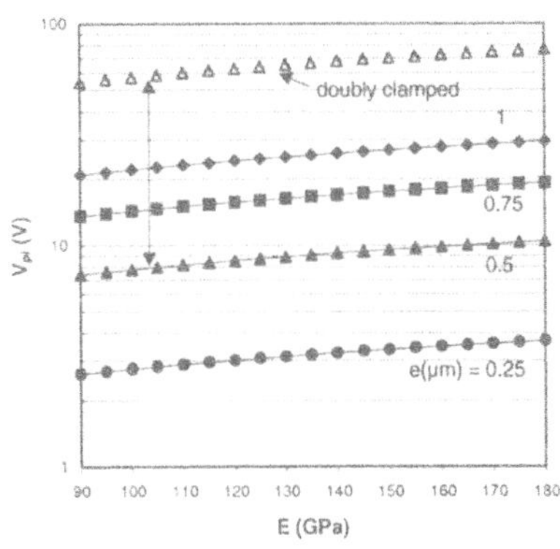

Fig. 3 Pull-in voltage, V_{pi}, as a function of Young modulus, E, with the inter-distance microcantilever-substrate as a parameter.

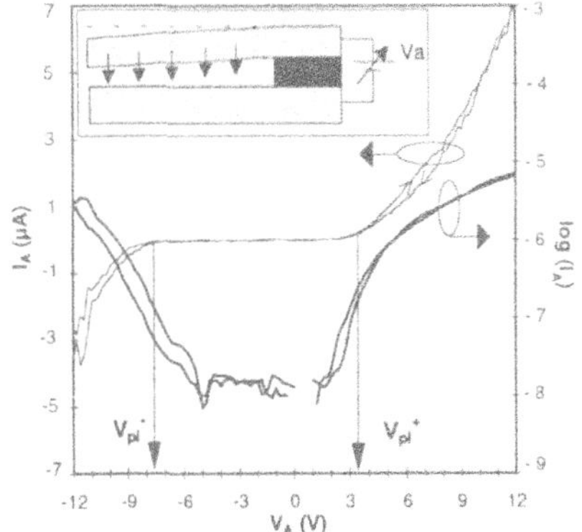

Fig. 4 Static characteristics, I_A-V_A, in linear and log scales (case of a polysilicon cantilever with L=80µm, W=12µm, e=0.5µm).

6 Practical work schedule

The microstructure fabrication starts in the clean room of CIME (Inter-university Center of Microelectronics), Grenoble, with groups of 4 students. Prior wafer processing allows to perform structure fabrication and characterization in 8 hours, as following:

- *fabrication and physical characterization* (4 hours). The work is started with two pre-processed wafers (Si_3N_4 and polysilicon already deposited). The students perform lithography of microcantilevers (first wafer) and microbridges (second wafer) followed by plasma etching. Then, the wet etching (HF) of SiO_2 is started. During and after etching, physical characterization measurements (optical microscopy, alpha-step, ellipsometry, 4-point probe measurements) are used as control steps. A critical point, which is under optimization, concerns the reduction of stiction phenomena. At the end SEM is used in order to verify the free-standing structures and measure the related dimensions. Various applications of free-standing structures realized by surface micromachining are discussed.
- *2D simulation and electrical characterization (4 hours).* SILVACO's ATHENA 2D numerical simulator is used during the first hour, as presented in Section 3. Afterwards, the HP4155A parameter analyser is introduced to the students and pull-in voltage measurements are conducted directly on wafers (using point probes). The last hour is dedicated to the caracterization of n-channel MOSFETs fabricated with the same technology and same masks as cantilevers.

7 Conclusion and further developments

As conceived, these practical works serve as an efficient introduction to MEMS fabrication technology and their first-order electrical characterization. Surface micromachining seems to be the most attractive solution for the fabrication of simple free-standing polysilicon structures, because of its total compatibility with the MOS technology. Further developments of these practical works will concern dynamic characterization and application of simple solutions to avoid stiction phenomena.

8 References

1. S. Wolf and R. Tauber, *Silicon Processing for the VLSI Era*, Lattice Press, Sunset Beach, California, 1986.
2. J.W. Gardner, *Microsensors: Principles and Applications*, John Wiley & Sons, Chichester, 1994.
3. S.M. Sze, *Semiconductor Sensors*, John Wiley & Sons, New-York, 1994.
4. S.D. Senturia, *Sensors and Actuators*, **4** (1983) 507.
5. H. Guckel, D.W. Burns, *Rec. of IEEE Int. Electron Dev. Meeting* (1986) 176.
6. H. Matoba, C.-J. Kim and R.S. Muller, *Proc. of 4th Int. Conf. Micro Electro, Opt, Mechanical Systems and Components*, Berlin, Germany (1994) 1005.
7. A.M. Ionescu, N. Mathieu and J.M. Terrot, *Proc. of the European Workshop on Microel. Ed.*, Grenoble, France, Ed. by World Scientific, Singapore (1996) 65.
8. N. Tar, T. Sonnenberg, H. Jansen, R. Legtenberg and M. Elwenspoek, *J. Micromech. Microeng.*, **6** (1996) 385.
9. R.K. Gupta, *Electrostatic Pull-in Test Structure design for IN-SITU Mechanical Property Measurements of MEMS*, PhD thesis, Massachusetts Institute of Technology, 1997.

EDUCATION ON COMPLEX MICROELECTRONIC SYSTEMS. HARDWARE - SOFTWARE CO-DESIGN WITH VHDL

E. SOTO, M.A. PEREIRA & S. FERNÁNDEZ
Dpto. Tecnología Electrónica. Universidad de Vigo
Apdo. Oficial. 36200-Vigo (Pontevedra) SPAIN
{esoto, sfdez}@uvigo.es

1. Introduction

Education on microelectronics is usually carried out with a bottom-up [1] approach. Basic concepts are introduced first, proceeding with the building of more complex systems later. A student has to take several different courses before starting to design more complex systems. Large modern systems are made up of many different components. Software has become tightly bound to hardware in these systems, therefore their design requires special skills in system integration and hardware-software co-design. The education on these skills are usually not addressed in the existing courses of electronic engineering. Hardware Description Languages are powerful tools to describe and to simulatevery large and complex systems [2].

In this paper we describe a new approach to the education of the design of very large systems using VHDL. In section 2 some background on the problems encountered when designing large systems is presented. Sections3 and 4 address the backbone elements of the proposed methodology. Section 5 shows the course flow-chart and discusses some alternatives. Finally in section 6 the conclusions are presented.

2. Complex systems

In the last years, education on the design of electronic circuits has become a difficult task because of the quick delevopment of new techniques and tools. Traditional education is focused on basic concepts and simple systems, providing good insights only in some particular aspects of the design. However, this approach is not useful to face the most modern systems.

Modern systems have grown too complex for an engineer to have a full command of all the specific topics related to them. Figure 1 shows a simplified PC board. Every circuit in that board is complex enough to be worth an engineering course. The problems arise from the design of the full board, a complex system which harbours components inside as complex as the system itself. Traditional education does not provide the students with the required skills to design this kind of systems, because of the long time it would require to analyze every component in the design.

T.J. Mouthaan and C. Salm (eds.), Microelectronics Education, 95-98.

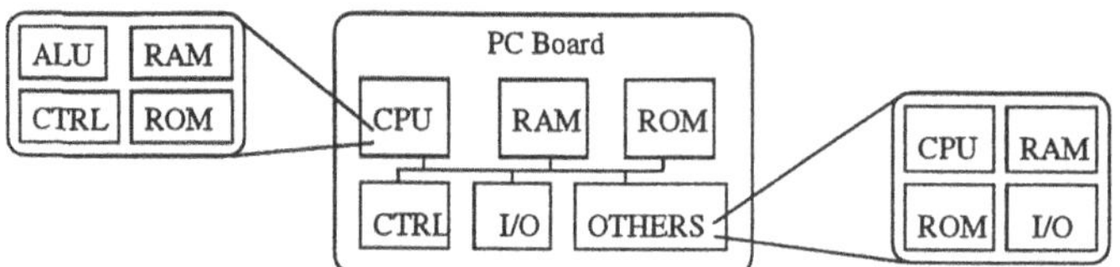

Figure 1: Simple PC board

Future engineers must be able to deal with complex system design in an efficient way. To succeed in this challenge, educators should provide their students with the suitable methodologies and tools.

3. Hardware-software co-design

System designers are not hardware engineers any more. Software and programmable circuits have become a fundamental part of systems. Software usually deals with the configuration, user interface, and complex tasks in a system. Hardware is aimed to basic and time consuming tasks that must be performed in real-time. Several different skills are required to deal with this: hardware design, programming languages and co-simulation techniques.

It is obvious that an engineer needs a long trainning to get knowledgeable enough to succeed in this kind of design. Partition of the design in small pieces is essential. But this partition fails in addressing the most fundamental assumption of system design: the fully assembled system must work.

Therefore, education on how to make hardware/software co-design is necessary for the next generations of electronic engineers. The question on how to perform this education has only been started to be answered in the last years. The advance of HDLs has allowed the development of courses that study different methodogies for complex systems design.

4. Microelectronic design with VHDL

VHDL is a very useful language to describe digital systems. It is a standard language since 1987 (last revision is 1076-1993).This character, and the increasing support in synthesis tools, have made VHDL an everyday tool in engineering design. It allows several levels of descriptions, from the simple interconection of components to highly abstract representations of system behaviour. The division into description levels varies with the author, but from a practical design point of view they can be classified in the following: gate, dataflow and behavioural.

While gate and dataflow levels are mainly used to generate synthesizable descriptions, behavioural is intended as a proof of concepts. Nevertheless, behavioural descriptions have several characteristics, that make them suitable for the education in the design of complex systems: Description of very large and complex systems with a few lines of code, Logic simulation of the described systems and Hardware-software co-simulations

Students skilled in the use of VHDL, can devote themselves to system level design strategies, software integration and test of large systems. Furthermore, provided suitable hardware tools, it is possible to perform hardware-software co-simulations to test the VHDL descriptions in a real life hardware environment.

5. Course flow

A short course is proposed, in order to achieve the following objectives: Systems design, Hardware-software co-design and Systems analysis and simulation. The course design example is showed in figure 2. The greater part of the system is provided to the student as VHDL descriptions (CPU, RAM, I/O blocks). There are freely available descriptions in VHDL of CPUs, like the DLX from [3] used in the recent Design contest [4] of the VIUF Fall'97 conference, in which the idea for this this course is based. Students design an ASIC and modify the description of the shadowed blocks to meet the design specificacions (ROM and control logic). The main goals of the course are: VHDL design of an ASIC, Processor programmation, System Integration and System Test.

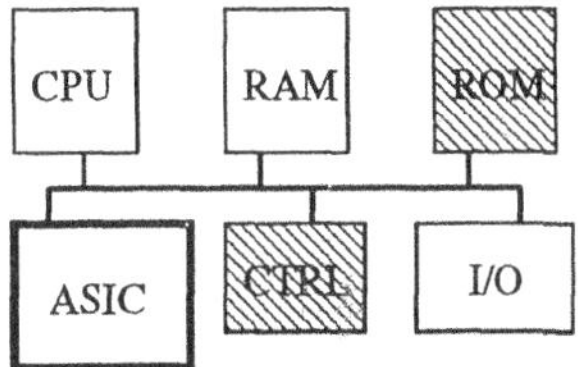

Figure 2: System design problem

As a first step the students should analyze the design problem. Design constitutes the next step. VHDL is used for description of the ASIC, from a behavioural point of view. Assembler or a high level language for the processor in which the system is based is used to program the ROM of the board. Suitable software to convert the source to a ROM description in VHDL is provided (all this is readily available for the DLX processor mentioned before).

Finally, test of the full system is performed. The students are encouraged to develop skills in finding the best way to test the system. A test proposal must be prepared, and a test bench created.

6. Conclusions

This paper shows different problems related to the design of large complex

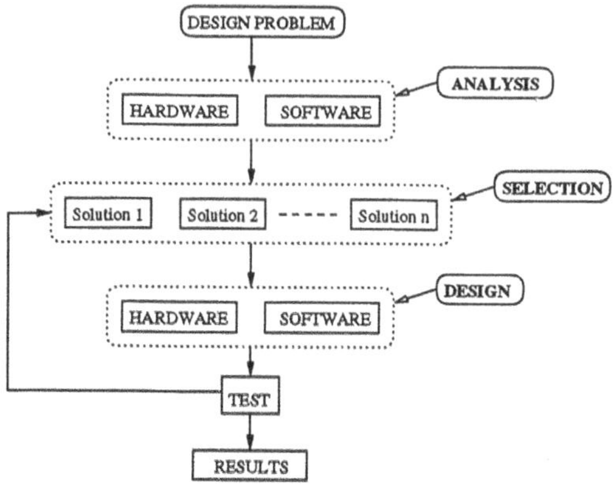

Figure 3: Course flow

systems. A short course is presented to provide the students with the proper background to deal with the design of complex systems, involving mixed hardware and software. as a intensive design course. The development of skills in analyzing the problem, searching for the best solution and planning of the test bench of the system are the main objectives of the course.

7. Acknowledgments

We thank to the VHDL International User Forum (VIUF) for the 1997 Design Contest Award, which this work addresses in part. Also to the Center for Reliable Computing (Stanford University) and the *Fundación PROVIGO* for their partly support.

8. References

[1] E. Hörbst, editor. *Logic Design and Simulation*, North Holland/Elsevier Science Publishers, 1986.

[2] Allen Dewey. *Analysis and design of digital systems with VHDL*, PWS Publishing Company, 1997.

[3] M. Gumm. *VHDL-Modelling and Synthesis of the DLXS RISC Processor*, University of Stuttgart, Institute of Parallel and Distributed High-Perfomance Systems, Dept. of Integrated Systems Engineering, 1995.

[4] E. Soto, M. A. Pereira and S. Fernández-Gómez. *VIUF Contest. Final Report*, Internal document, Dpto. Tec. Electrónica, Universidad de Vigo, 1997.

REALIZATION IN CLEAN ROOM AND ELECTRICAL CHARACTERIZATION OF P-TYPE THIN FILM TRANSISTOR

Prof. O. BONNAUD, D. GUILLET, Prof. F. RAOULT, Dr A.C. SALAUN
Groupe de Microélectronique et Visualisation,
Centre Commun de Microélectronique de l'Ouest,
Campus de Beaulieu, Université de Rennes 1, 35042 Rennes Cedex
France
tel: +33 (0) 2 99 28 60 71 fax: +33 (0) 2 99 28 16 74
e-mail: bonnaud@univ-rennes1.fr

1 Abstract

This technological work is intended to DESS students (bac+5 level) in the frame of a project included in their cursus. The main goal is to apply the microelectronics technology knowledge already got after one week spent to fabricate a monocrystalline MOS structure in the cleanroom of the Centre Commun de Microelectronique de l'Ouest (CCMO). After this preparing phase, the students improve their technological know-how and skills by fabricating thin film transistor (TFT) with new process steps such as undoped or in-situ doped polysilicon deposition and reactive ion etching. The total duration of this project is about 80 hours, a major part occuring in the cleanroom of CCMO.

2 Introduction

This technological work is intended to DESS students (bac+5 level) in the frame of a project included in their cursus. The main goal is to apply the microelectronics technology knowledge already got after one week spent to fabricate a monocrystalline MOS structure in the cleanroom of the Centre Commun de Microelectronique de l'Ouest (CCMO), which is one of the centers of the french committee for education in microelectronics (Comité National pour la Formation en Microélectronique or CNFM). During this preparing phase, the students use mainly the usual process steps such as oxidation, photolithography, wet etching, doping diffusion and aluminium deposition steps. The fact of fabricating thin film transistor (TFT) allows to improve their skills and knowledge in new process steps such as undoped or *in-situ* doped polysilicon deposition and reactive ion etching. Note that this type of TFT's could be included in a large area electronic circuit for instance a flat panel displays or a projection matrix if they are fabricated on glass substrate. The total duration of this project is about 80 hours, a major part occuring in the cleanroom of CCMO.

T.J. Mouthaan and C. Salm (eds.), Microelectronics Education, 99-102.

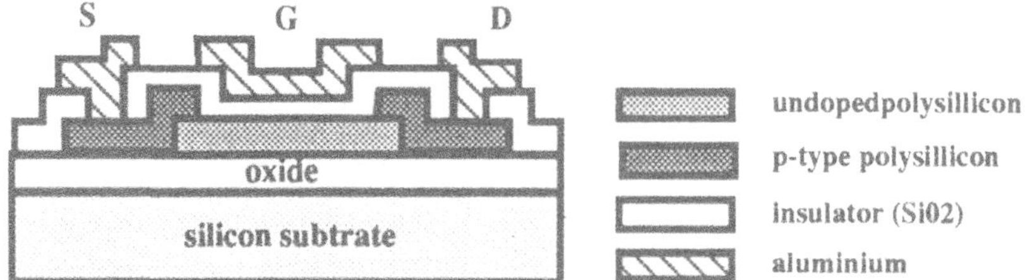

Figure 1 : Cross-section of a polycrystalline silison thin film transistor. This technology is called planar beaceuse the active layer is in the same plan that source and drain ones

3 Experiment

TFTs are elaborated through a four mask process. A cross-section of the device is shown figure 1. Polycrystalline layers are obtained by low pressure chemical vapor deposition (LPCVD) at 550°C using silane at a pressure fixed at 90 Pa. At this temperature, the layers are amorphous deposited. The crystallization occurs either during a post annealing or during a high temperature step such as oxidation. Previous works have shown that the best results in term of electrical behaviour of the TFTs are obtained when the crystallization occurs during the oxidation step [1]. Thus, a first 150nm thick un-intentionnally doped amorphous silicon layer is deposited on oxidized monocrystalline substrate and etched by dry SF6 plasma to define the active layer (channel region). A second 150nm thick boron doped amorphous silicon layer (p-type polycrystalline silicon layer) is also deposited by Low Pressure Chemical Vapor Deposition (LPCVD) technique, but in this case, a dopant gas, diborane, is added to the silane. This *in-situ* doped LPCVD step is not so easy than the undoped deposition due to the fact that the boron atoms exalt the deposition rate [2], and make difficult the obtention of homogeneity of the layer thickness along the reactor. This *in-situ* doped layer is then etched to form source and drain regions. After RCA cleaning, a 60nm thick gate oxide is obtained by thermal oxidation at 950°C for 2 hours and followed by a photolithography to define the gate. Finally, aluminium is deposited and wet etched to define source, drain and gate contacts.

To control the gate oxide quality, capacitors are fabricated on monocrystalline substrates and the oxidation made simultaneously to gate insulator one. These capacitors have front and back aluminium contacts ; a good flat band voltage indicates a good quality of the dry oxide. The gate oxide thickness is controlled by ellipsometry measurements and checked from the accumulation mode capacitance-voltage measurmeents. Note that this four mask process needs more than one equivalent full week of work for the students.

4 Electrical characterization

TFTs were electrically characterized with a HP 4145 semiconductor parameter analyser, connected to a prober set inside a black box to prevent any light influence on the electrical behaviour. To check the electrical behaviour of the fabricated stuctures, direct and reverse characteristics have to be performed. The output

characteristics, plotted in linear coordinates gives informations on the mobility of the channel. The characteristics plotted in semi-logarithmic coordinates allows to evluate the subthreshold slope and the reverse current. These two parameters give informations on the density of states at the oxide-channel interface and on the quality of the polycrystalline layer respectively. This approach constitutes a good training on physics of semiconductor course.
The output characteristics is plotted figure 2. It shows a good behaviour with clear linear and saturation modes, and a good modulation of the drain-source current with the gate voltage.

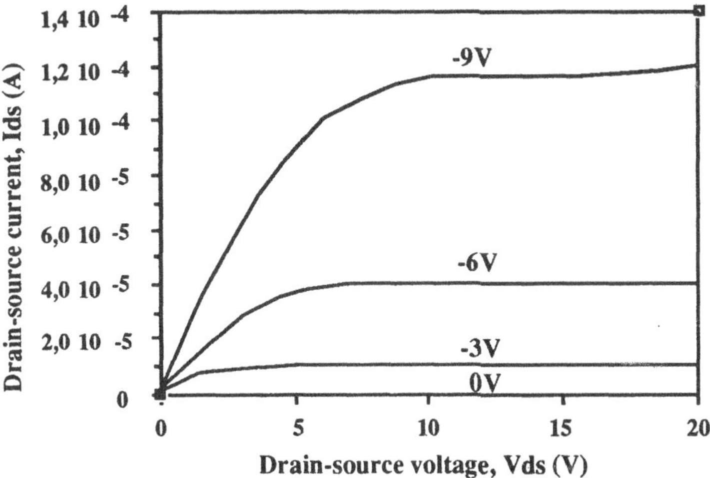

Figure 2 : Output characteristics of the p-type thin film transistor.. Linear and saturation modes are clearly displayed. The gate voltage well modulates the source-drain current. (Vgs = 0, -3 -6, -9 V)

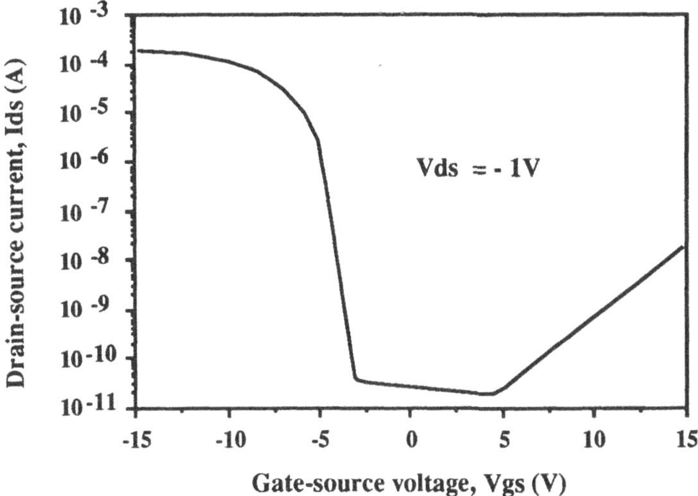

Figure 3: Transfer characteristics of the p-type thin fil transistor. An acceptable reverse current is observed as well as a good subthreshold slope and a good Ion/Ioff ratio.

The transfer characteristics plotted for $|V_{DS}| = 1$ Volt and shown figure 3, gives evidence of an acceptable reverse current and a good subthreshold slope. Let us note that this features are importnant for potential applications. The Ion/Ioff ratio at $|V_{GS}| = 10V$ is higher than 10^5, that is rather good for this type of transistors.

5 Conclusion

As a conclusion, this pedagogical experience shows that it is possible to process some unusual structures with graduate students. Of course, the developped process is a little bit long and needs the use of specific step such as *in-situ* doped LPCVD technique which is hard to control. But, this technique takes more and more importance in the integrated circuit foundries, especially in submicron CMOS technology. This work constitues a good experience for the students.
Thanks to the success of this first experiment with graduate students, this new process developed in the CCMO cleanroom should be proposed for usual technological trainings in the future.

References

[1] A.C. Salaun, A.C. S, B. Fortin, T. Mohammed-Brahim, K. Ki-Sion, L. Haji and O. Bonnaud, Thin Polycrystalline silicon films annealed at 950°C : structural and electrical properties and application to thin film transistors, Polycrystalline Semiconductors IV - Solid State Phenomena, vols. **51-52**, 1996, pp. 603-608

[2] D. Briand, M. Sarret, P. Duverneuil, T. Mohammed-Brahim, K. Kis-Sion, Influence of the doping gas on the acial uniformity of the growth rate and the electrical properties of LPCVD in-situ doped polysilicon layers, *EURO CVD 10, Venise (Italiy) 10-15 Sept. 1995.*

Session C

NEW CONCEPTS IN TEACHING

MICRO ELECTRONICS SYSTEMS DESIGN

A Professional Micro Electronics Education

Ir. W.J.Th. De KAPER.
MESO project manager, The Hague University of Professional Education
Johanna Westerdijkplein 75, 2521 EN The Hague, The Netherlands

1 Abstract

Described here is an original way of teaching, founded almost completely on carrying out projects for companies. Students start working on these projects in the second semester of their first year, with is unusually quick. The special kind of organisation needed to make this possible is discussed here. In the execution of the projects, there is no difference with professional design companies. A conscientious description of specs and deadlines is mandatory.

2 History

In recent years, there were several serious indications that besides the regular education for engineers in digital and analog hardware and software, there was a need for more than one kind of engineer.
In the past, we focussed (in our traditional education in electronics) very much on pure technical/ scientific knowledge, with which we educated students to be engineers very able to proceed their study on a Technical University or to accept a job at a major company in Microelectronics.
However, a lot of medium-sized companies don't design the (high-tech) products they sell themselves; they purchase the products, but high quality support is included. These kind of companies do need engineers, with technical knowledge (without the skill to design and create products) AND management (science) skills. So, some years ago, the Hague University created an education "Commercial Engineer".
But the social importance of small and medium-sized enterprises which do develop their own, often quite sophisticated products has increased with the years, and those enterprises need a kind of engineer, capable of product development. Long periods to break a person in are unacceptable. So, the student has to experience product development during his period of training; for this purpose, we created the MESO education.

At the Hague University, we believe to satisfy, in these ways, all demands of the society:
- A 'classical' education : proceeding of study / major companies in Micro-electronics;
- A 'commercial' education: for engineers at companies which do not develop the high-tech products they sell.
- The MESO education : for engineers at small and medium-sized companies which do develop their own products.

Besides that, many students considered a study in Micro-Electronics to be 'dull and difficult', resulting in lower and lower numbers of students. An opinion poll among students showed the reason: the student was capable of using his formulas, but designing a working machine was still impossible. In one way or another, all those reasons pointed to the lack of practice-experience. So, how can we get genuine practice in the education-program ? How can teachers support several teams consisting of just a few students, how can we make it affordable ? How can we get orders for projects?. This is a story about the many 'how's' we were confronted with, and the ways we found to solve them, some good, some less good.

T.J. Mouthaan and C. Salm (eds.), Microelectronics Education, 105-108.

3 Shock of Culture

In the last section we mentioned the lack of experience in developing end-products. In magazines there is often some talk of a 'shock of culture' the student/engineer suffers from. If we want to prevent this, we will first have to investigate and to verbalise the reasons:

- High order of regularity, exact timetables at school, versus the 'chaos' at a company; there is always something urgent and unexpected crossing the task in question. The inexperienced engineer suddenly has to face a quite dynamic environment.
- By means of prelims and exams the student was tested. The way the young engineer is tested, is the practice. In the past, there was always a close attention not to ask about a subject unknown to the student. The young engineer will have to face daily unknown subjects; not only in the field of microelectronics. As an engineer, he should be capable of handling these 'unknowns'.
- It is impossible to provide the student with all the knowledge he will ever need. So, he has to find a way to fill in the knowledge he is wanting in.
- In the - more or less - isolated environment of a school, the student never has to deal with and to look for suppliers for the parts he needs. Once at a company, these luxury is over.
- During his study the student was not only told how to do it, but also what to do. The young engineer will be told what to do also, but in a much more global sense.
- As a student, most designs were on paper and ended with a computer simulation. As an engineer, this is just the first phase. Sometimes - an exception - a laboratory model was build. As an engineer, after the laboratory model comes the prototype, then the series-0, finally the production version. How should the inexperienced engineer know how to do all this? It's the first time he has to take full account of safety requirements, EMC, sturdiness, protection against water, requirements of pick-and-place machines, etc.
- The electronics and software will be only a part of the product. It must be mounted in some case, and important decisions has to be taken : metal, plastic, supplier, price, safety, etc. The engineer has to be familiar with the terms of the supplier.
- At the company, the student has to face: planning, timetables, deadlines, (resource) management, budgets, allocation of tasks, delivery time, meetings with the customer, presentations, dependence on other departments, etc etc.

4 Weekly diary

To give practice in all the aspects of the job of the engineer, is not possible without practice itself. It will be not enough to specify and to develop one's 'own' products. Out-of-school companies are needed, to bring into the walls of the school (1) the unlooked for (2) pressure of deadlines and budget. Before a short overview of the way MESO is organised, we will sum up some of the questions and problems which came to the desk in a special MESO week, to see the degree the MESO education approaches the later practice of the engineer.

- ...appointment with Andre du Croix from a telecommunications company to discuss a project about IP telephony...
- ...order to develop an industrial, EX IIa, industrial Windows CE machine with dedicated maintenance software. In a week, the customer wants to give a demonstration at an important seminar; we need urgently the Windows CE 2.0 upgrade; we need to connect a big VGA screen with a small Windows CE organiser; a PC-card is still not available, and now ?.
- ...good morning. We have a problem. We have an important order here to deliver in the near East, but inside we use a PCB from French origin and now that company is bankrupt. Can you make it for us in three weeks ?...
- ...the PCB, format A3, for the Test Adapter we developed, is received. Now the 2000 golden nails has to be mounted. How ?....
- ...why can't Intel provide the BSDL (Boundary Scan Description Language) file for there own ethernet controller which has hardware BST facilities inside ?....
- ....we need five pieces of this part. It can be delivered in quantities of 250 or more....
- ....framing a contract for an in-company course about Boundary Scan Technology....
- ....a little party with the students of the project team at the national telecom company (PTT). Four months ago they supplied us with an order for the development of a new kind of calculator for the costs of telephone conversations. The laboratory model was ready in two

months, just in time. Now, four months later, the first 1000 pieces are delivered to the users, again just in time.

- ....the Doppler system to measure the speed of water shows bad noise margins, outside the specs; how can it be improved ?
- ....can you design for us a dynamic traffic control for Istanbul ?....
- ....yes we want your order to design a vision system for plant cuttings (but where can I purchase some expertise ?)....
- ....yes we want your order to design a 10 GHz navigation radar system (but where to find facilities, equipment and expertise ?)...

5 MESO organisation

The education is divided in semesters.
The first semester is the only period the student has nothing to do with contract-orders. In the first 6 weeks, he has to do 3 learning projects of 2 weeks each. The purpose of these learning projects is to become familiar with his design environment. The first project concerns chip design and digital systems, the second analogue systems, the third is about programming. It isn't necessary to understand everything, the main purpose is practising within the design environment (So please don't bother the teacher with FAQ's about your tools *after* this period !)
For the remaining part of the first semester, there is one subject for every week. After a two-hours lecture, the student has six hours a day to work on his project of study, defined by the teacher.
In the second, third and fourth semester, the mapping of the day is : a lecture of two hours, working on his project of study for three hours, working on his project of business for the remaining three hours.
The fifth semester the student is doing work experience, in the sixth semester he graduates. Only one of these two periods is outside school; the other period is inside school and the student has, so near at the end of his study, the responsible position of project manager. His (often 4) team members for instance are two students from the second and two from the fourth semester. In designing his masterpiece (of course this an contract-order of some company) he has to prove his engineering capabilities, but that's not enough. He also has to prove his *managing capabilities*. He has a team to lead. He has to negotiate with customers and suppliers, do presentations and keep an eye on the planning and budget. Lacking facilities must be made up. *So, can he get the job done ??*
We haven't said to much about the teacher in his new role. Nevertheless, he is still very essential. But he will have two different tasks now : at first the classical one of being an expert in his speciality, at second as a tutor, who looks at, watches over, facilitates, supports and criticises the process that is going on. He will be tutor for some projects and maybe specialist for other projects; the degree of participation in the fundamental design is in the first instance controlled by the degree of difficulty of the project; and so are the costs.

The acquisition is done mainly by the teachers. In the course of years, with hard labour, a network of contacts with companies is realised, and this together with a list of successful projects, is the foundation of the acquisition. Within the walls of the university there is a special agency, Technomarkt, that is the corporate body in the contracts. Concerning the acquisition, the daily practice is the following :

- The teacher receives a request, or approaches a company by his own initiative;
- A first pilot meeting to check for possible co-operation;
- A second meeting, together with the specialist-teacher, estimates of attainability, duration and costs;
- A visit of a the Technomarkt canvasser at the company;
- Submit a quotation and fix up the contract.

6 Business projects

In this section we will give terse descriptions of some major business projects.

- In close co-operation with JTAG Technologies we developed an embedded 386-based computer board, fully boundary scan testable, as a target to demonstrate all the vector generating and assembly faults diagnosing possibilities of the JTAG software. This A5 sized 6-layer PCB with flash memories and in-system programmable logic is suited for development of 386 systems and it can also be used to show the spectrum of sound on a LCD display. As an alternative, the JTAG logo can be placed on the LCD by means of the boundary scan chain. A second project was a Test Adapter. If a brand-new component with BST facilities is introduced on the market, a BSDL file is supplied with it. Sometimes such a BST file has errors, with the Test Adapter it is possible to check for it. Besides these two, JTAG gives also order to design two other projects.
- One of the biggest and most difficult projects was to develop complete and certified software for chipper and chipknip ('plastic money'). We designed also the hardware, just to be able to test the software, because the third party could not deliver the hardware in time.
- Simple and low cost hardware (included software) for controlling and monitoring mains plugs in large buildings, from a central office, is an ever repeated request. We have designed three systems with complete different characteristics.
- The telephony costs calculator is a PIC microcontroller based machine. During phoning, it calculates the costs by the second, not by 'blips'. So the calculator has to know the rates by day and by night, and it should also be aware if a call is local or not, etc.
- We have designed a cruise-control for Rhine barges, a GPS application, to make it possible for such ships to arrive just in time, to save expenses for fuel.
- When a fire-brigade with high-pressure air enters a burning house, the colleagues at the fire engine must be alarmed when a certain time is passed. We designed an electronic alarm system for this purpose.
- To warm the glass houses (to cultivate vegetables and flowers) large furnaces are build. The electronics to make the right gas/air mixture is designed by MESO.

7 Conclusions

- To design a quite professional electronic product is a strong driving force for teacher and student.
- The distance between teacher and student disappears for a large part, both have the same goal.
- Regular revenues from the Department of Education are insufficient for this kind of teaching, companies will have to pay a realistic amount of money to make it possible. Because the teaching will be done in small teams, with a lot of support; modern equipment with the newest software and expensive licences are needed.
- Students which do not have the talents to design electronic products, are confronted with 'hard technics' in an impressing way; the selection process will be quite natural. The student doesn't need to be convinced by the teacher that he has chosen the wrong profession, because he will find out this himself.
- Teachers are no longer all-knowing persons; they are just the persons to take the lead in looking for the places and persons with the knowledge needed.
- Intensive co-operation with a number of companies (some with a lot knowledge of electronics, some without) changes the static environment of a school in a quite dynamic scenery.
- By no means it will be possible for the limited number of teachers to do all the work. The energy of the last phase students must be brought into action.
- In the individual assessment of the students, the task of the teacher is as difficult as the task of his colleague at some company, who has to make judgements about his subordinates. At this moment, we are working very hard to make good and objective assessment criteria for the kind of project education described before. But what do you want ? A very relevant but not so accurate assessment, or a not so relevant but very accurate one ? We choose the first, and then we try to make it as perfect as possible.
- Education in microelectronics can be most inspiring, under the condition that students can work on the solution of real world problems. We witness students complaining about the closing of the laboratories at 21.00; we witness students working in holidays. Please, teach us more about.......

THE GM/ID METHODOLOGY ON MATLAB:

a pedagogical tool for analog integrated circuit design education

D. FLANDRE and P. JESPERS
Microelectronics Laboratory, Université catholique de Louvain,
Place du Levant 3, B-1348 Louvain-la-Neuve, Belgium
tel: +32.10.478135; fax: +32.10.472598; e-mail: flandre@dice.ucl.ac.be

1 Introduction

Analog integrated circuit design is usually a difficult matter to teach and learn since it generally requires expertise in both device and circuit behaviour and modelling. In this paper, we describe a promising education experience which has been developing at UCL for a few years. The idea is to couple and implement on MATLAB, the results of a symbolic analysis of the analog circuit and a representation of the device behaviour which is strongly related to the expected performance of the circuit and does not depend on the device dimensions or bias voltages which are part of the unknowns of the design procedure.

2 Basic g_m/I_D methodology

The key element in our design methodology is the transconductance-over-drain current characteristic (i.e. g_m/I_D) versus normalized current $I_D/(W/L)$ (fig. 1). Typically the MOS g_m/I_D shows a maximum plateau in weak inversion followed by a monotonic decrease in moderate and strong inversion. This curve can easily be fitted by a continuous MOS model such as the EKV model [1].

Consider then the example of the single saturated MOS OTA stage represented in figure 2. Its performance are related to the transconductance g_m, the drain current I_D and the Early voltage V_A assuming that the output conductance g_d be given by I_D/V_A. Once a pair of values out of g_m/I_D, g_m or I_D has been derived from the OTA specifications, the device aspect ratio W/L is unambiguously determined from the monotonic g_m/I_D vs normalized current curve (fig. 3). This can be achieved using either a model like the EKV or the experimental characteristic of the device itself, covering thus all modes of operation of MOS devices. The result can also finally be unambiguously related to the OTA dc operating point, i.e. gate bias, saturation voltage...

T.J. Mouthaan and C. Salm (eds.), Microelectronics Education, 109-112.

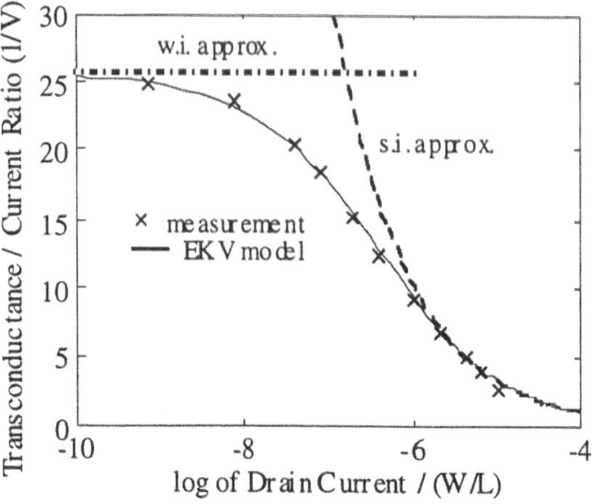

Fig. 1 : Bulk n-MOSFET g_m/I_D characteristic.

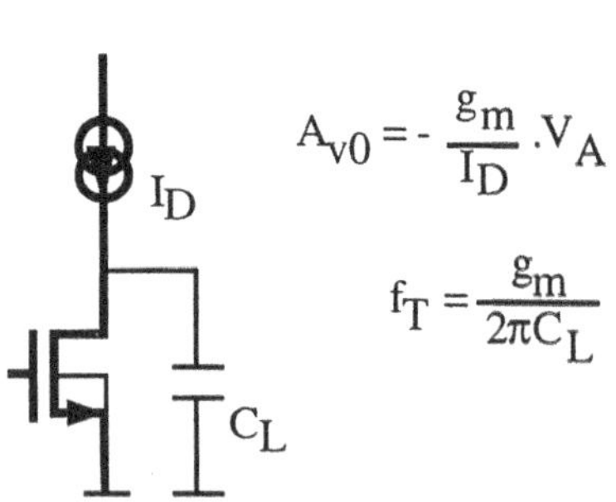

Fig. 2 : Intrinsic MOS gain stage and dc open-loop gain and transition frequency expressions.

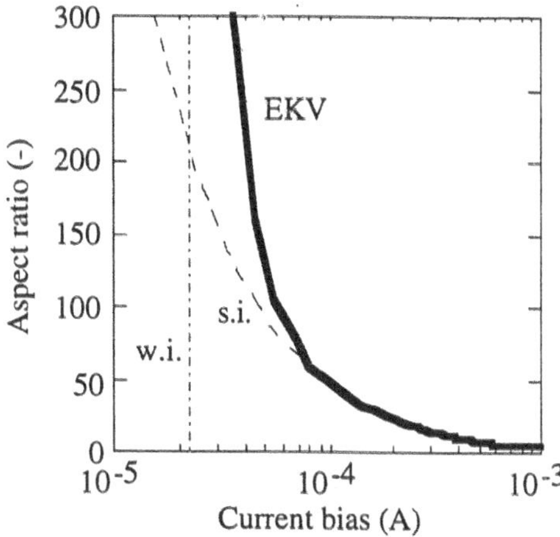

Fig. 3: (W/L) vs I_D locus for 10 MHz @ 10 pF, comparing EKV, strong or weak inversion models.

Our methodology is then closer to a top-down approach, starting from the circuit specifications towards the device level unknowns (size, bias, current ...), whereas conventional SPICE approaches need to fix good initial guesses for the device level unknowns in order to start iterative simulations which may require a lot of experience. The general design methodology is summarized in figure 4 and can be easily implemented and solved on MATLAB, allowing the students to quickly test various initial choices and check their influence on the circuit synthesis results, performance, limitations ...

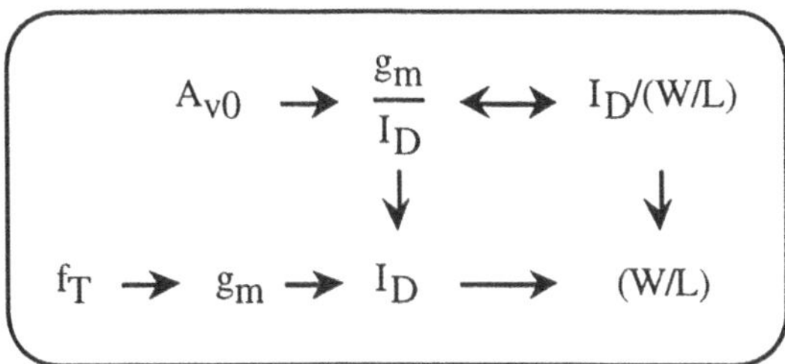

Fig. 4: Systematic g_m/I_D design procedure for the MOS intrinsic gain stage of fig. 2.

3 Extension to complex OTA architectures

The actual synthesis procedure has been successfully extended to a number of CMOS OTA architectures (single stage OTA with and without cascoded output stage [2, 3]; Miller [4] and folded-cascode 2-stage OTA [5]...), considering a variety of small- and large-signal specifications (gain, transition frequency, phase margin, settling time, noise, slew rate, distortion...), and operating conditions (micropower [2, 4]; high-temperature [3]; high-frequency [5]...) as well as applications. The synthesis procedure has been implemented in a number of student exercices and projects and its validity successfully demonstrated by the experimental realizations of several of those circuits.

The potential of the technique can be illustrated by the synthesis of a 2-stage Miller CMOS OTA for a given f_T-C_L specification taking into account that the device transconductances, sizes and feedback capacitance C_f have to be carefully optimized for stability considerations. Classical first-order small-signal analysis yields that :
- the DC open-loop gain A_{v0} and transition unity-gain frequency f_T are given by

$$A_{v0} = \frac{g_{m1} \cdot g_{m2}}{g_{d1} \cdot g_{d2}} = \frac{g_{m1}}{I_{D1}} \cdot \frac{g_{m2}}{I_{D2}} \cdot V_{A1} \cdot V_{A2} \quad \text{and} \quad f_T = \frac{g_{m1}}{C_f} \qquad (1),$$

where the suffixes 1 and 2 refer to the active devices of the differential pair and of the output stage respectively.
- for stability considerations, i.e. to get a phase margin of 60° at f_T, the following conditions must be met:

$$g_{m2} = 10. g_{m1} \quad \text{and} \quad C_f = 0.11. \left[C_1 + C_2 + \sqrt{(C_1 + C_2)^2 + \frac{4}{0.22} . C_1 . C_2} \right] \qquad (2),$$

where C_1 and C_2 are the total device and parasitic capacitances at nodes 1 and 2 which can be straigthforwardly estimated from the device sizes.

Combining this set of equations on Matlab, with the "g_m/I_D vs $I_D/(W/L)$" relationship and the required technological data (i.e. only the body effect factor, the V_A parameter and the source/drain and gate capacitances per unit area), we can extend the top-down strategy of fig. 4 to calculate, for given f_T - C_L - phase margin (ϕ_m) specifications and as a function of the gain (i.e. the g_m/I_D and lengthes of the devices), the actual problem unknowns (i.e. the device widths and bias currents). Figure 5 illustrates some results of such a process, obtained in just a few seconds of CPU time, showing the actual limitations of a given technology, e.g. here the optimal current consumption, corresponding to moderate inversion design, large device widths and high gain, above which the device sizes tend to be too large and hamper stability.

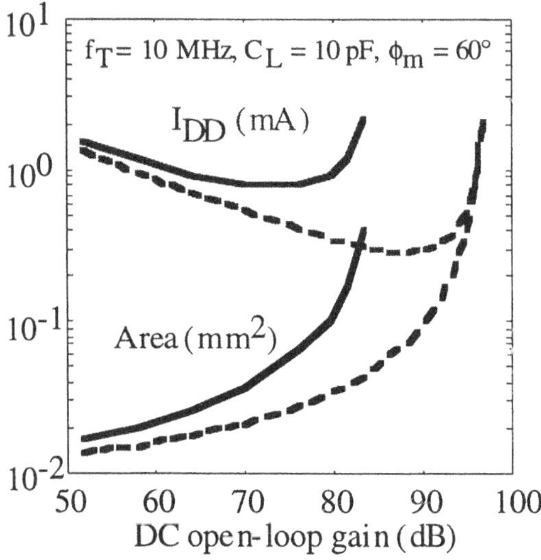

Fig. 5: Miller OTA total supply current and die area performance limitation vs Av0 for bulk (solid lines) or SOI (dashed lines) 2 μm-CMOS technology.

4 Conclusion

We have developed a systematic approach for designing CMOS OTAs with optimal performance based on the combination on MATLAB of symbolic analysis with a g_m/I_D representation of the MOS behavior. The efficiency of the technique both for education and actual design has been demonstrated by a number of exercices and examples.

5 References

[1] C. Enz *et al*, Analog Integrated Circuit and Signal Processing, vol. 8, pp. 83-114, 1995.

[2] F. Silveira *et al*, IEEE Journal of Solid-State Circuits, vol. 31, pp. 1314-1319, 1996.

[3] J.-P. Eggermont *et al*, IEEE Journal of Solid-State Circuits, vol. 31, pp. 179-186, 1996.

[4] D. Flandre *et al*, Solid-State Electronics, vol. 39, pp. 455-460, 1996.

[5] D. Flandre *et al*, IEEE Journal of Solid-State Circuits, vol. 32, pp. 1006-1012, 1997.

HOW TO MAKE THE STUDY OF MICROELECTRONICS MORE ATTRACTIVE

D. DONOVAL
Department of Microelectronics, Slovak University of Technology
Ilkovicova 3, 812 19 Bratislava, Slovakia

1 Abstract

The microelectronics education system is under strong pressure to change. Its infrastructure must reflect the dynamic technological changes connected with the increasing complexity of integrated circuits and microsystems. To maintain the innovative force of microelectronics in the future, highly qualified experts with a multidisciplinary background and experimental skills must be educated [1].
The microelectronics education is generally considered to be a difficult study programme. Implementation of advanced simulators for the process, device, circuits and system level design allows a unique insight into the structure operation, thus contributes to overcome the inadequate interest of students in microelectronics. Intensive collaboration and dissemination of courses and modules based on CAL (Computer Aided Learning) tools through the Internet/WWW environment is of great importance not only for inter-university collaboration but also for time and cost-effective retraining and lifelong learning of engineers and technicians from industry.

2 Introduction

The domination of electronic industry over other industrial branches [2] and increasing spread of applications of microelectronics and microsystems is not adequately reflected by the interest of students in the study of microelectronics. This can be attributed also to the required physical and chemical background and corresponding mathematical treatment which labels the microelectronics education as a non-easy task. Therefore, undergraduate students at many universities are educated at best as designers or technicians, versed only on using recipes, rather than also innovative engineers and scientists [3]. The users can design their circuits and verify their electrical properties using advanced software tools without understanding the basic physical principles. To overcome the aversion of students in theoretical study advanced simulators based on numerical modelling and simulation of technological processes, electronic devices and circuits play an increasingly important role [4,5,6].

T.J. Mouthaan and C. Salm (eds.), Microelectronics Education, 113-116.

3 Our approach

The classical bottom-up way of teaching starts with materials and structure properties continues with device and circuits fabrication and principles of operation towards system level design. Our approach is a top-down approach. It is based on the implementation of different simulators in educational process. The students start with the design of electronic systems and circuits using advanced design tools with large libraries of subcircuits, modules and devices as black boxes. Then they verify the electrical properties of the designed circuit using the circuit simulator. Circuit simulation evokes further questions related to the electro-physical properties of individual components. To analyse the influence of device parameters from material and structural points of view other software packages such as parameters extraction tools, device and process simulators must be used (Fig. 1). Individual simulators provide users with answers to their questions, which are connected with a proper function of the designed circuit or its individual components. The analysis of the influence of temperature, frequency and applied voltages in a wide range can be made easily. The multiple window demonstration allows simultaneous correlation of output electrical characteristics with the internal structure properties and thus provides users with a unique insight into the internal operation of semiconductor structures [7]. In general, modelling and simulation contribute considerably to an easier way of understanding the complex principles. Thus the subject – design and analysis of electronic systems and circuits based on basic physical and technological principles – may become of greater interest and may attract more students to microelectronics.

The better understanding of the physical and chemical background is strengthened by the increasing complexity of the fabrication process due to a smaller feature size and the use of new materials, increasing complexity of the electronic systems, circuits, and microsystems due to their higher integration and utilisation of new principles [8].

The quality of the software tools from the educational point of view is of great importance in this process. The simulators must have an interactive user friendly graphical interface for data visualisation. They should have modular architecture allowing the introduction or substitution of new and more complex physical models for the old ones. The use of industry-standard simulators with some modifications for the educational process allows teaching a lot of practical stuff and gives students hand-on experience. By implementing them into practical training they provide a very nice environment for problem oriented learning to discover an "own new phenomena".

The globalisation of the microelectronic industry stimulates collaboration between universities and industry on national and international levels. The preparation of CAL courses based on different commercially distributed design tools and simulators (e.g. by Cadence, Viewlogic, TMA, and ISE) for educational purposes is of great importance. The creation of targeted hypertexts using the Java language [9] allows to use the simulators as a core tool for animation of a corresponding text which contains a lot of "nothing saying" equations. Its distribution by the Internet/WWW features facilitates for individual practical training either for university students or for continuos lifelong learning and/or retraining for engineers and technicians from the industry.

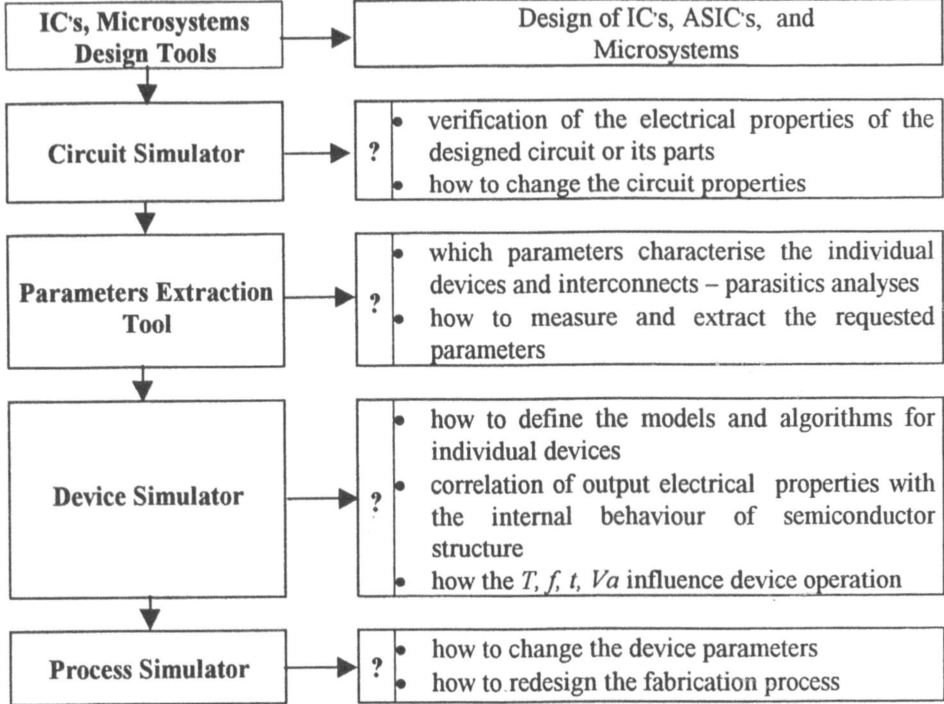

Fig. 1. The inverse use of modelling and simulation in microelectronics education

4 Conclusions

The microelectronics education infrastructure is under intense pressure to change. The origin of this big demand can be summarised in the following items:

- dynamic technology which approaches the physical and technological limits
- higher complexity of designed integrated circuits and microsystems
- inadequate R&D government funding to universities, especially in CEE countries
- livelong training and retraining of engineers and technicians from the industry.

To fulfil these demands, great challenges have to be solved by universities in the modification of study programmes and methods of teaching, particularly

- the compulsory fundamental courses must by multidisciplinary to provide the students with the necessary background, e.g., the designers of VLSI circuits with millions of gates should have such fundamental knowledge like the physics of the transistor. The main deal is which material should be dropped and which added.
- in spite of that, the study programme must be relatively open and flexible with a set of well defined optional courses and/or modules which allow specialisation.

- a problem-oriented project must be an inevitable part of the study programme at least from the 3rd year of study to enhance the students skills in selected field.

CAL base courses can contribute to this effort considerably, especially

- they provide the learners with an easier and better understanding of complex problems using simulation to animation of "boring" textbooks and thus attract more students to study microelectronics.
- they provide a time- and cost-effective environment for practical training of users via solving problem-oriented projects.
- they can be disseminated easily through the Internet/WWW environment, thus being accessible not only to students but also to engineers and technicians from the industry.
- the worldwide dissemination of courses supports either inter-university collaboration or the collaboration with industrial partners on the national and international levels and contributes to the problemless mobility of students, which is very important for globalisation of microelectronics industry.

Although the above mentioned items are promising to modify and update the microelectronics curriculum they are still serious restrictions which limit the innovation process. They can be summarised as

- inadequate motivation of teachers for the modification of current study programmes and/or in preparation of hypertext based courses and their dissemination through Internet/www environment.
- inadequate collaboration between the universities to prepare and disseminate the courses on the basis, which is best for individual institutions, programme. To avoid the current situation where every school tries to cover everything.
- inadequate collaboration with industry in general and local levels to specify and fulfil their actual needs and requirements.

Acknowledgement - This work was supported by grant No. 95/5195/286 of the Slovak Ministry of Education.

References

[1] What's Next for Microelectronics Education? A D&T Roundtable, IEEE Design&Test of Computers, Vol. 13, p. 95, 1997

[2] COURTOIS, B., CAD and Testing of IC's and Systems, Where are we going?, TIMA Grenoble 1994

[3] SAH, C. T., Fundamentals of Solid State Electronics, World Scientific, Singapore 1991

[4] HICKS, P.J., DAGLESS, E., JONES, P., KINNIMENT, D., LIDGEY, F., MASSARA, R., TAYLOR, D. and WALCZOVSKI, L., EDEC-A New Courseware Package for Electronic Design Education, In: Proceedings of 3rd Int. Conference on Computer Aided Engineering Education, p.33, Bratislava (1995)

[5] WACHUTKA, G. and VOIGT, P., Computer Aided Training in Design and Fabrication of Microdevices and Systems, In: Proceedings of 3rd International Conference on CAEE, p. 345, Bratislava (1995)

[6] DONOVAL, D., BEDLEK, M., BARUS, M., RACKO, J. and HROMCOVÁ, J., DEVSIM, A. New Simulator for Better Understanding of Internal Semiconductor Structure Behaviour, Comput. Mat. Sci., 1, p.51, 1992

[7] DONOVAL, D. RACKO, J., WACHUTKA, G. and BALTES, H., A Multiple Window Demonstration of the Operation of PN Diodes Using DEVSIM, Int. J. Engn. Ed.,11, p.36, 1995

[8] KAMARINOS, G., For a New Educational Strategy for ULSI Microelectronics, Material Science and Engineering A 199, p. 45, 1995

[9] Homepage for Javasoft's, a Sun Microsystems Company, Mtn. View, California, Hot Java, http:/www.JavaSoft.com/HotJava/

TRENDS ON MICROELECTRONIC SYSTEMS EDUCATION

I. C. Teixeira, J. P. Teixeira
IST (Instituto Superior Técnico), Universidade Técnica de Lisboa, Av. Rovisco Pais, 1096 Lisboa Codex, Portugal

Abstract

The world moves now in a context of accelerated change, complexity and uncertainty. As a result of its mission, as a locomotive of scientific advance and human resources qualification, academia faces new challenges, particularly in the area of Microelectronic-Based Systems (MBS) education. In order to cope with the technologic state-of-the-art as well as its trends, classic courses on microelectronics must be complemented with hardware/software co-design course(s), and with reliability and testing course(s) , in order to satisfy present product Quality requirements. Moreover, technical, economical and organizational fundamentals need to be considered, namely Concurrent Engineering. Finally, continuing education courses (using multimedia) are key enablers, to provide professionals the ability to develop the skills that allow them to master the factors of change.

1. Introduction

Sir Winston Churchill once said: "The empire of the future is the empire of the mind" (as quoted in [1]). The present world context shows the veracity of these words. In fact, professionals are more and more forced to selectively acquire and master new information and knowledge and to continuously upgrade their knowledge and skills, in order to undertake and keep rewarding jobs. This requires, first, that formal education (in Portugal, a 5-year graduate program), must qualify engineers with the fundamentals to support (in a long-term basis) their professional lives. Second, continuing education, should be envisioned as an enabler to the continuous learning process [2].

In the engineering field, this involves constant updates on technologies, methodologies and tools, and new organizational and working methods [3,4]. More specifically, in MBS education, key issues must be addressed and solved: the fast pace of evolution of supporting technologies (electronics, communication and information), ever increasing complexity and interdisciplinarity, team work, ever decreasing time-to-market and ever increasing customer requirements [5,6]. The purpose of this paper is to present the contribution of the authors to take into account this trend, and to describe its implementation on MBS education at IST.

The paper is organized as follows. Section 2 briefly describes the engineering educational trends in the information society, and its impact on new course profiles, exceeding technical expertise. Section 3 presents some of the courses offered in our 5-

T.J. Mouthaan and C. Salm (eds.), Microelectronics Education, 117-120.

year graduate course, 2-year M.Sc. course and continuing education program. Finally, section 4 summarizes the main conclusions of the work.

2. Engineering Trends

As business organizations move from Task-Centered (TCO) to Process-Centered (PCO) Organizations, people work on *processes*, not on individual tasks. A *process* is a complete, end-to-end, set of related tasks (or activities) that together create value for a customer. Well designed processes are now much simpler than before, due to task interaction and integration. Conversely, the activity each person has to perform becomes more and more demanding, as they become professionals. Therefore, in PCOs, complexity shifts from processes to jobs.

A *professional* is now understood as "someone who is responsible for achieving a result, rather than for performing a task" [3]. Professionals working together share a common goal – make the process successful. They act as part of a process *team* encompassing people from marketing, design, production, after sales support, accounting, etc. At any timeframe, the engineer's stronger connections are to one or more business processes, rather than to an Engineering Department as before. Such new context quests for *new career profiles* [2], and for continuous learning.

As the context around academia rapidly (and continuously) changes, so must educational programs adapt to changing requirements [7]. However, Universities should not to become strict business corporations, being individual qualification its main "product". Modern concepts of process optimimization should not kill the university culture. Therefore, pragmatism and prompt intervention in the market needs to be conciliated with creativity [8], deep reasoning and analysis.

3. System Engineering Courses

As knowledge base continuously increases, and application areas for MBS increase (Fig. 1), there is a need to resist to the tendency of trying to teach more and more in the same timeframe. We refer as *Microelectronic-Based Systems (MBS)*, and not just Microelectronic Systems, the hardware/software systems that use, as dominant physical support, microelectronics, (although, other technologies, such as optics and mechanics can be used, like in microsystems). Hence, MBS engineering is no longer viewed as a *hardware* engineering discipline. In fact, the 'term' is applied to a *hardware/software* (hw/sw) discipline, as MBS products and MBS-based services may include networks of hw/sw MBS units, interacting with information systems.

More specifically for the 5-year graduate program (Licenciatura) on Electrical and Computer Engineering [6], our experiment is based on the following approach. For the MBS related education, a set of mandatory courses first includes the hw and sw fundamentals (electromagnetism, circuit theory, solid-state electronics, circuits and systems, languages, algorithms, sw engineering, etc). In our view, MBS engineering must take into account *the whole product development* cycle [9]. In order to grasp the technical expertise, students need to acquire the fundamentals of three main areas (fig. 2), namely, *circuit and system design techniques, Electronic Design Automation (EDA) methodologies and tools, and manufacturing technologies.*

Furthermore, the fundamentals not only of problem solving, but also of problem *preventing*, need to included. This significantly broadens the specification and design sub-processes, as quality, reliability and DFX [9] clearly modulates system requirements, design and performance. DFX means Design For X, where X stands for implicit requirements, or quality criteria, such as Testability (DFT), Manufacturability (DFM), or Environment (DFE) [10].

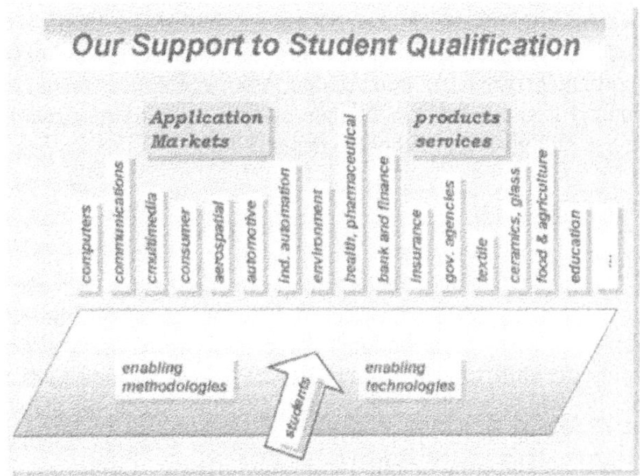

Fig. 1 – Application areas for MBS engineering.

As a consequence, a mandatory introductory course on microelectronics and integrated circuit design, that we refer as ***ISE (Integrated Systems Electronics)***, is included in the program curricula.This course is intented to provide technical, hands-on expertise, using commercial EDA tools, and also awareness to the above mentioned issues.

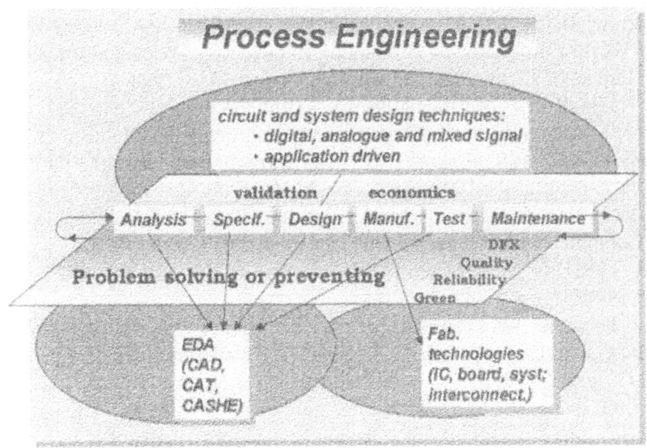

Fig. 2 – Main areas of MBS engineering.

Additionally, a set of *optional* semester courses are offered. Here, two main options can be chosen. First, technical courses on analogue integrated circuit design, microwave MBS design, microsystems, or test and reliability can be selected. Second, two courses are being offered: ***HwSwCo (Hardware/Software Codesign)*** and ***ESE (Electronic***

Systems Engineering). Concepts included in these courses prove to be very useful for the final year project, usually carried out by multidisciplinary teams working on a concurrent engineering environment. Advanced topics of these areas are also offered at M.Sc. level. Course curricula can be found at our home page (http://zorba.inesc.pt). Also, MBS continuing education is being launched [2], especially using multimedia courses, namely [11] and [12]. An alpha version of the latter is available at http://webnt.ist.utl.pt/ebse. In the ESE course, Concurrent Engineering [4] is considered, and product non-functional requirements (e.g., quality, testability, dependability, safety and environment) are taken into account. Also, a process-centered Product Development environment is simulated, to stimulate multidisciplinary team work, design reusability, and the use of high-level methodologies and tools.

4. Conclusions

In this paper, engineering trends have been highlighted, in order to show how they influence university MBS education. As shown, selective technical education needs to be complemented by organizational and economical education. Also, classic hardware focus must be shift to a hardware/software focus. Moreover, design courses need to consider the whole product development process. Hence, we showed our contriburtion for implementing it at IST. ISE, Electronic Systems Engineering (to consider Concurrent Engineering) and Hardware/Software Co-design courses are briefly described, for formal education (5-year program, and M.Sc.). For continuing education, multimedia courses, one of which available at the web for distance learning, are referred.

5. References

1. C. Rose, M.J. Nicholl, "Accelerated Learning for the 21th Century - the 6-Step Plan to Unlock your Master-Mind", Delacorte Press, 1997.
2. I.C. Teixeira, J.P. Teixeira, M. Pile, D. Durão, "From Continuing Education to Continuing Learning Using Self Assessment and Process Monitoring", accepted for presentation at the 7th World Conference on Continuing Engineering Education: the Knowledge Revolution – the Impact of Technology on Learning, Torino, May, 1998.
3. M. Hammer, "Beyond Reengineering - How the Process-Centered Organization is Changing our Work and our Lives", HarperBusiness, HarperCollins Publishers, Inc., 1996.
4. D.E. Carter, B.S. Baker, "CE (Concurrent Engineering): the Product Development Environment for the 1990s", Addison-Wesley Pub. Co., 1992.
5. "What's Next for Microelectronics Education?", a D&T Roundtable, IEEE Design and Test of Computers, vol. 14, pp. 95-102, Oct.-Dec., 1997.
6. E.A. Lee, D.G. Messerschmitt, "Engineering an Education for the Future", IEEE Computer, vol. 31, nº. 1, pp. 77-85, Jan., 1998.
7. I.C. Teixeira, J.P. Teixeira, M.Pilc, D. Durão, "Quality Development in Engineering Education at IST", Proc. SEFI Seminar, "A Tool to Improve the Learning Process", June 19-21, Grimstad, 1997.
8. E. Lumsdaine, M. Lumsdaine, "Creative Problem Solving: Thinking Skill for a Changing World", McGrawHill Int. Ed., General Engineering Series, 1995.
9. AT&T Technical Journal, vol. 73, No.2, Special Issue on "New Electronic Test Technologies", March/April 1994.
10. T.E. Graedel, B.R. Allenby, J.C. Sekutowski, "Green Product Design", AT&T Technical Journal, Vol. 74, Nº.2, pp. 17-25, Nov./Dec. 1995.
11. J. P. Teixeira, "Microelectronic Systems Technology", Computer-Based Training Course, Continuing Electrical Engineers Education (CEEE-CBT) Project, EUROFORM, IST/OE/FUNDETEC and Hogeschool Utrech, Euraltis, Academic Medical Centre, 1994.
12. I.C. Teixeira, and J.P. Teixeira, "Trends in Electronic-Based Systems Engineering", Self-Training Project, December, 1997.

Session P3

MULTIMEDIA IN MICROELECTRONICS EDUCATION

MULTIMEDIA WEB-BASED COURSEWARE ON MICROELECTRONICS

D. CAVIGLIA, G. DA BORMIDA, D. PONTA, M. TERRIZZANO, M. VALLE
Department of Biophysical and Electronic Engineering
University of Genoa
Via Opera Pia 11A I-16145 Genoa - Italy
Phone +39-10-3532759, fax +39-10-3532175
E-mail ponta@dibe.unige.it

1 Abstract

This paper outlines the structure and methodology underlying a Web-Based Course (WBC) on Microelectronics targeted to the students of the third year of the Electronics Engineering curriculum. The contents are delivered through custom-made courseware, based on a hypertext and integrated by tools for practice and evaluation purposes. The use of multimedia features and interactive models of basic components and circuits enhances learning. Together with pedagogical and technical issues, the paper discusses the pedagogical effectiveness of a Web-based environment integrating learning material delivery and teacher-learners interaction, using properly designed tools.

2 Introduction: the learning environment

With the widespread and growing diffusion of Internet connection, the Web is nowadays becoming a quite popular way to deliver not only information but also learning material and, in some cases, complete courses. Web-Based Courseware (WBC) is therefore taking its place in the world of new pedagogical methods and systems besides other technologies, such as Computer-Based Learning (CBL) and different implementations of distance learning [1].
The vast majority of current WBC realizations are based on HTML pages integrated by images being therefore equivalent to a traditional book. On-line delivery provides advantages, such as the possibility of easily updating the material and the hypertextual structure, a naturally embedded feature of Web sites. Nevertheless, from the pedagogical pointy of view, such material is no better than traditional books and it does not take advantage of computer resources to help the learning process.
Our experience has shown us that students do not like to read hypertexts on the screen, but still prefer traditional books [2]. The new learning material needs to take advantage of the computer not only for the technical aspect of information delivery, but also for

T.J. Mouthaan and C. Salm (eds.), Microelectronics Education, 123-126.

pedagogical purposes. In short, we need to build a learning environment where students find resources to help understanding concepts, to evaluate progresses and to do practical work [3].

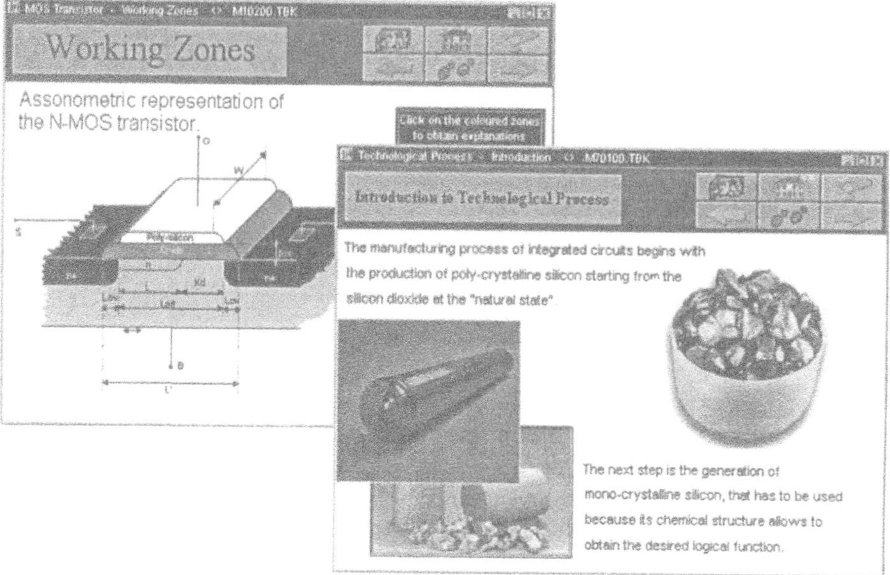

Figure 1 *- This figure represents two typical situations of the courseware. While the left screen-shot shows an example of 3D interactive model of the N-MOS transistor, the right one is a multimedia hypertext page.*

3 Course methodologies and their implementations

Our group has been engaged since a few years ago in the development of Computer Based Learning (CBL) material to be employed as a support for the teaching of fundamental issues of Analog and Digital Electronics. The project has been supported in the past by the European program COMETT (project WORKBENCH) and now by the Telematics Application Program (project ARIADNE) [4]. The unifying aspect of the work is the constant effort to develop an environment where the basic pedagogical activities are enhanced by the use of computers and software tools [5, 6].

The learning environment, originally designed for introductory courses in Analog and Digital Electronics, lends itself very well to be applied to Microelectronics. It is divided into four parts, representing the pedagogical activities upon which the course is based [6]. The presentation of theory, concepts and other information is the first, followed by the interactive explanation of the theoretical material just introduced. Verification of learning is achieved in the third one, while practice of synthesis and analysis is the target of the last one. Hypertextual material, implemented in Asymetrix ToolBook, includes pictures, 3D graphics, animations, pedagogical simulations and interactive learning tools to provide an intuitive idea of a concept, together with or instead of a textual explanation. The use of multimedia features, such as pedagogical MPEG sequences [7], and interactive 3D models of basic components and circuits enhances learning (fig. 1).

Verification features, usually as multiple-choice questionnaires, are scattered through the course. Introduction to real design issues and professional CAD tools is achieved with a custom-built hypertext interface to the widely adopted PSPICE simulator, that allows learners to test digital and analog schematics of their own design (fig. 2).

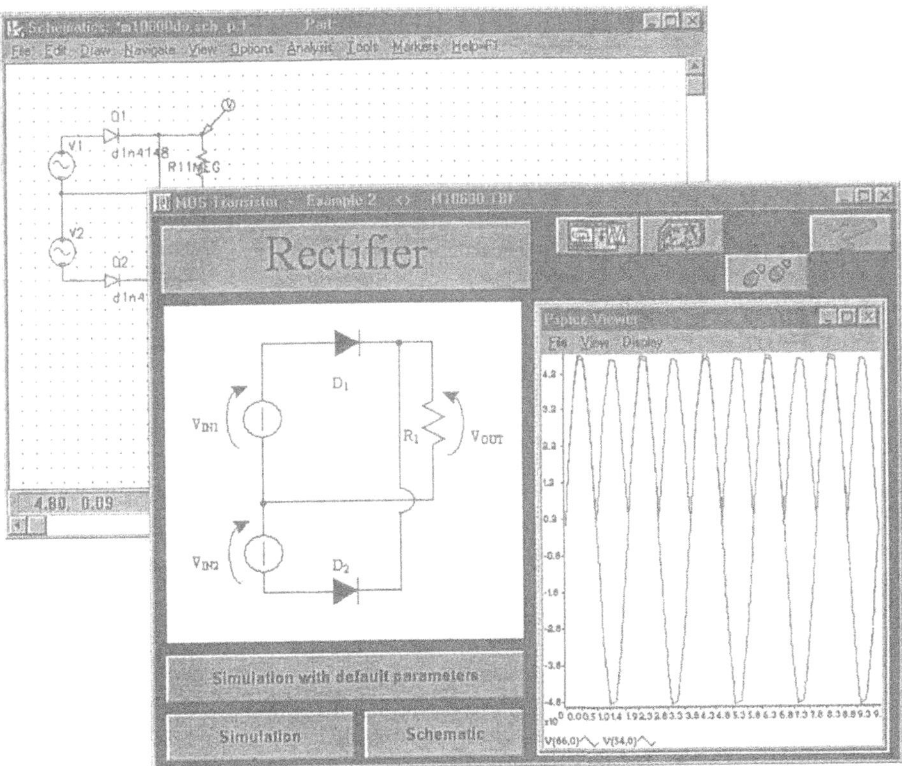

Figure 2 - The hypertext interface to the PSPICE simulator (in background) has been custom-built to allow learners to test digital and analog schematics of their own design.

One of the main targets of the work is to make the learning material suitable for delivery using a Web-based environment, where a different organization is necessary to optimize network resources. Network delivery implies on-line courseware consultation, reducing to a minimum the storage necessity at student's end and taking maximum advantage of a network server as a controlled and always up-to-date repository of courseware. Nevertheless the local situation of networking infrastructures, i. e. narrow bandwidth, does not allow a full implementation of this approach.

We have decided to use the Netscape Navigator browser to deliver ToolBook documents, configuring it to handle TBK files. The Web-based distribution of ToolBook documents, that contain multiple links, is not automatic, because the last version of ToolBook needs to have all the files available on the local machine. For this reason we have developed a Dynamic Link Library (NeTbk.dll) to interface ToolBook and Netscape. NeTbk.dll implements a Dynamic Data Exchange (DDE) server and client to transparently download TBK files when needed. The system is configured to automatically save in a local directory a copy of the files that have been used, in order to access them remotely only when an updated version is available and re-build step-by-step a personal copy of the courseware [8].

4 Pedagogical contents: the course on Microelectronics

The WBC (that will be also called in the following "courseware") in Microelectronics is a classical introductory course on microelectronics systems, devices and techniques, corresponding to the academic course with the same name. The course is addressed to the students of the third year of the Electronics Engineering curriculum of the Laurea degree, that lasts five years and fairly corresponds to the Master degree. The Microelectronics course, that introduces the fundamental aspects and technology of integrated circuits, is targeted to students with basic knowledge on electronic devices, circuit theory, digital systems theory (from the second year), analog electronics, computer science and automatic controls (from the third year). The course and the WBC on Microelectronics address the following main topics: MOS device characterization; MOS fabrication technology; basic MOS digital circuits and analog circuit stages; BiCMOS technology. Lectures are integrated by guided laboratory exercises that, concerning circuit simulation of CMOS logic gates and layout generation for a generic CMOS process, are based on a simple educational software [9] with 3D technology process simulator, on-line circuit simulator and simple layout editor.

5 References

1. T. C. Montgomerie, and D. Harapnuik, "*Observations on Web-Based Course Development and Delivery*," International Journal of Educational Telecommunications, vol. 3, no. 2/3, pp. 181-203, 1997.
2. G. Da Bormida, D. Ponta, and G. Donzellini, "*Teacher-learners cooperation produces an innovative computer-based course,*" in Proceedings of 1996 AACE Eighth World Conf. on Educational Multimedia and Hypermedia (ED-MEDIA), Boston, USA, June 17-22 1996.
3. G. Da Bormida, D. Ponta, and G. Donzellini, "*Learning Environment for Digital Electronics,*" Proceedings of 1996 IEEE Int. Conf. on Systems, Man and Cybernetics (special session on "Multimedia Learning Environments"), Beijing, China, October 1996, vol. 4, pp. 2892-2897.
4. Project ARIADNE Programme: Alliance of Remote Instructional Authoring and Distribution Networks for Europe (ET 1002, European Union, DG XIII, 4FWRP, Telematics for Education and Training), 1996.
5. P. J. Mosterman, J. O. Campbell, A. J. Brodersen, and J. R. Bournem, "*Design and Implementation of an Electronics Laboratory Simulator*," IEEE Trans. on Education, vol. 39, no. 3, pp. 309-313, August 1996.
6. G. Da Bormida, D. Ponta, and G. Donzellini, "*Methodologies and tools for learning electronics,*" IEEE Trans. on Education - Special CD-ROM Edition, vol. 40, no. 4, pp. 291, November 1997.
7. V. Bhaskaran, and K. Konstantinides, "*Image and video compression standards* - Algorithms and Architectures" Second Edition, Kluwer Academic Publishers, Norwell, USA, 1997.
8. D. Ponta, and G. Da Bormida, "*Re-engineering a Computer-Based Learning Course in Digital Electronics or Flexibility, Re-use and Network Delivery,*" in Proceedings of 1996 IEEE-ASEE Frontiers In Education Conf. (FIE), Salt Lake City, USA, November 6-9 1996.
9. E. Sicard, "*Introduction to Microelectronics*," Project Manual V4.1, INSA, Toulouse.

REMOTE EDUCATION EXPERIENCE ON LEARNING IC CHARACTERIZATION/PRODUCTION TEST

Y. BERTRAND, R. LORIVAL, M. ROBERT and G. CAMBON
Laboratoire d'Informatique de Robotique et de Microélectronique de Montpellier (LIRMM), UMR CNRS/Université Montpellier II no 55060 161, rue Ada 34392 MONTPELLIER Cedex 5 France.

1. Introduction

Electronics manufacturers are driven to produce new integrated circuits (IC's) with higher quality in shorter production cycles (the well known time-to-market concept). In parallel, with the advent of submicron technologies digital circuits become more complex, with an increasing transistor-per-pin ratio making them hard-to-test without Design-For-Test (DFT) and Built-In Self-Test (BIST) facilities [1]. Also, during the past decade the production of mixed analog-digital IC's have been dramatically developed due to the explosive growth of the multimedia and telecom markets and testing for quality assurance has proven to approach 50% of the manufacturing costs for some complex mixed-signal IC's [2].
In this context the test requirements have to be considered as soon as possible in the design flow. Thus the electronic engineers might have to face some test problems during their design phase. Also, due to (i) the very fast technological evolution of Automatic Test Equipment (ATE), and (ii) the advent of new types of circuits (mixed digital/analog circuits) and new test techniques (boundary scan, I_{ddq} monitoring) the test engineers have to use new testers and solve new test problems. Indeed there is a real need for initial and continuing education in the field of automatic testing. Semiconductor manufacturers ask for electronic engineers having full competencies in IC testing in order to (i) guarantee the quality of the circuits over high production throughput and (ii) reduce the test time and then the test cost (purchase and maintenance costs).

2. The CRTC: a common center for test

To meet the industrial requirements for test and product engineers the CNFM (Comité National de Formation en Microélectronique) decided to create a common test service in its Montpellier center (Figure 1). So, from late 1997 the CRTC (Centre de Ressources de Test du CNFM) is hosted by LIRMM (Laboratoire d'Informatique, de Robotique et de Microélectronique de Montpellier) and benefits from the environment

T.J. Mouthaan and C. Salm (eds.), Microelectronics Education, 127-130.

of an internationally renowned test team (10 permanents researchers with an annual average production of 20 international papers).
In order to train a maximum number of students with a reduced cost the choice was made to have a unique ATE reachable by net. Under these conditions, a week of training in the common test center of Montpellier may be replaced by an adaptable period of training (distributed in time or localized during a single week, or any other combination) in each local center. This permits to save money for the displacement of, say 20 students per CNFM center. Of course the price to pay is the perfect maintenance of the test tool all along the year and a good coordination between students and their educators in the local site on one side, and technical people in the common test center on the other side. Organizing a test training which aims at educating 20 students in parallel (using 10 workstations) and which is based on the activation of real physical tests in a remote test center reachable by net is not a very common challenge.

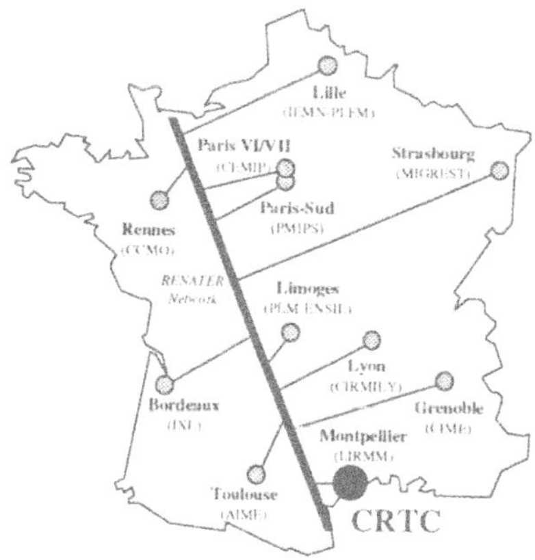

Figure 1 : The RENATER network connection of the 11 CNFM sites

3. Training organization

The first technical constraint comes from the test tool itself: due to the physical aspect of the test operation, ATE does not admit several users in parallel. This implies the different working groups have to execute their test operations sequentially. Remark this constraint is not specific to remote education, it also exists in a context of a local one. Indeed, this situation is not really critical considering that the major part of the time needed to develop a test program may be spent without any connection to the tester. The tester itself is controlled by a dedicated workstation (Figure 2), the so-called «test station», which is equipped with all the software tools needed to develop complete test programs. As long as we develop a test program without the necessity of applying it on the circuit, it is enough to work on the test station only, in the so-called «off-line mode». The only moment we have to connect the tester («on-line mode») is during the test operation itself. The duration of this test operation is very short (classically a few seconds for the complete test of a standard circuit) and it is not a major problem to sequentially execute the various test operations several times for the different working groups. As the test tools implemented on the test station may be concurrently reached by several users and knowing that most of the time needed to develop a test program may be spent through an off-line connection it is not a problem to ensure the local training of 20 students in the common test center.

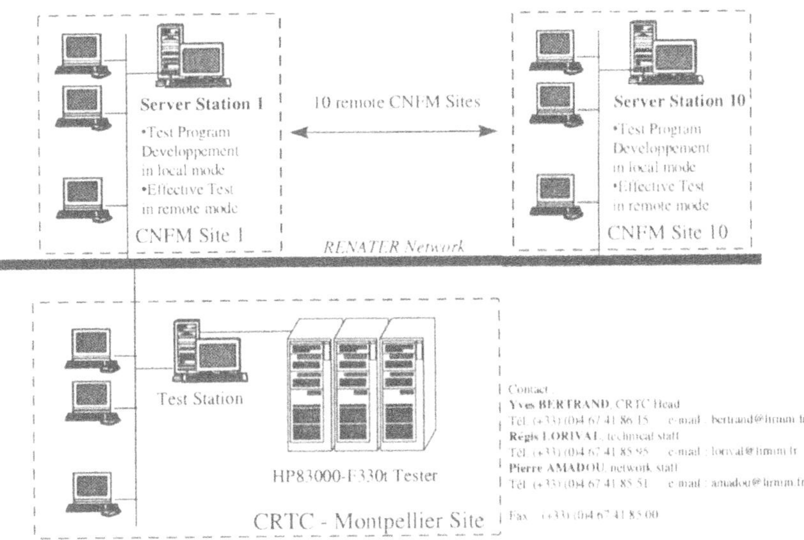

Figure 2: Network connection for remote education in IC testing

The next point concerns the way we can extend this configuration to a remote education. Taking into account the present-time performances of the networks we think it is not really practicable for 20 students to develop test programs in parallel using a network-dependent connection. Consequently it has been decided to equip each CNFM center with a local server hosting all the test resources needed for test program development (Figure 2). This server exactly acts at the local level as the test station does at the test center level. The test station only differs by the fact it is directly connected to the tester.

The inter-connecting structure chosen to implement the remote education for all the CNFM sites is shown on Figure 2. The common test center of Montpellier (CRTC) is equipped with all the hardware and software resources for test. The tester itself, the HP8300-F330t of Hewlett-Packard, is a high-performance tool designed for characterization and production testing. The test station is an HP9000/755 workstation which contains all the software resources for test programming and tester interfacing. This test station is directly connected to the tester through an ultra-fast optical link. Each local center is equipped with a HP9000 PA server station containing all the software resources for test programming. Up to ten workstations may be connected on this server thus allowing twenty students to concurrently prepare their test programs. When one of the ten working groups has finished its test program it establishes a remote connection to the CRTC test station. Then, using an on-line configuration, this group may reach the tester and launch the real physical test of. Of course this operation necessitates to have the good circuit at this moment on the test head. In the context of a training this constraint is not very strong due to the fact that all the groups work on the same circuit. Moreover in a very near future, we plan to acquire some video means to accompany this new educational experience. This would allow to somewhat compensate the miss of a real physical contact with the tester for the remote-logged students and to facilitate the interface between trainees in their local site and people from the test center.

4. Continuing education

The project of using network facilities to organize remote courses in IC testing was first decided in the context of initial education. Nevertheless it appears that the need of test courses in the context of continuing education was also very sharp. This comes from the fast changing situation in the field of IC testing. With the very rapid development of new mixed digital/analog circuits for telecommunication and multimedia markets there is a strong need of new dedicated methods and systems for test. A novel generation of mixed-signal digital/analog testers appears and there is a lack of test engineers able to efficiently manage them. The HP83000-F330t tester of the CRTC is one of these up-to-date testers with a fast increasing penetration curve in the test market. Taking this context into consideration the CNFM has decided to establish a partnership between CRTC and Hewlett-Packard to permit continuing education to be proposed to industrial partners. Several one-week training have been already organized in the CRTC to educate company engineers on these new test systems. Up to now these training have been directly organized in the test center of Montpellier. For the next months it is planned to extend these possibilities to some other CNFM centers.

5. References

1. M. Abramovici, M.A. Fridman and A.D. Brauer: Digital Systems Testing and Testable Design, IEEE Press, New-York, 1995
2. A. Grochowski, D. Bhattacharya, T.R. Viswanathan, K. Laker: *Integrated Circuit Testing for Quality Assurance in Manufacturing: History, Current Status, and Future Trends*, IEEE Transactions on Circuits and Systems-II Analog and Digital Signal Processing, Vol.4, No 8, pp. 610, August 1997.

HYPERMEDIA IN THE ANALOG INTEGRATED CIRCUIT DESIGN EDUCATION

A. FERREIRA, J.L. NOULLET, E. SICARD
Institut National des Sciences Appliquées
INSA - DGEI, Complexe Scientifique de Rangueil,
31077 Toulouse, France

1 Introduction

Hypermedia can be considered as the combination of hypertext and multimedia, two hot topics of the 90's.
A hypertext is just a text containing links to other documents, that may be local or reachable by means of some network, in order to enable the reader to obtain more details on any subject mentioned in the text, only if he needs. Hypertext is a real revolution in the manner of presenting information, and beyond the popular vision given by its application on the Internet, it is gradually replacing conventinal documentation in the industrial context, with the greatest impact in the field of micro-electronics : the major vendors of C.A.D. software have already stopped distributing paper manuals for their products. The same trend begins its progression in scientific publications, pioneered by the "IEEE Transactions on Education" [1] which proposes an annual issue made of hypertext only.
Multimedia is the art of combining in the same electronic document two or more of the following media : text, still images, animated images and sound. In addition, a decent multimedia implementation must provide some interactivity, the first degree of which being the hypertext links (i.e. hyperlinks). In the field of electronics education, we can observe two trends : in the first trend animated images and sound are used continuously, the multimedia document being comparable with a video tape, with the addition of fast navigation by hyperlinks [2], in the other one the skeleton of the document is the hypertext, no sound is involved (for classroom usage) and animated images are not considered as essential, even if there are some interesting exceptions [3].
Interactivity can be considered at two levels : a fine level, where the user can use the mouse for moving something on the document and observe what happens [4], and a global level where the user is invited to be a part of the system by reacting by producing himself some hypermedia stuff.

T.J. Mouthaan and C. Salm (eds.), Microelectronics Education, 131-134.

2 Strategy choices

The objective of our work was defined as development of hypermedia material or "courseware" for classroom use, in the context of practical integrated circuit design training.
The following guidelines were stated :
- attention must be caught on the interest of hypertext in industrial context
- global level interaction is strongly desired : the students are invited to report their results in the same form they received information
- the courseware must be portable, not depending on a defined computer type
- since training is made by "real experiment" [5] on professional C.A.D. tools, the courseware is not required to offer fine level interaction (but this is not excluded)
- maximal flexibility : the material has to be adjusted nearly in "real time"
- no sound, little or no animation.

For these reasons, the choice of the HTML language was obvious. It may be objected that specialized "authoring tools" allow to produce more sophisticated multimedia documents at a quality level which is becoming a standard in commercial CDROMs, but we did not want the courseware to look like a commercial CDROMs but rather to have nearly the same austerity as the on-line documentation of the professional C.A.D. software. This choice is enforced by some leading examples like the IEEE Transaction on Education [1] which is HTML-based, a silicon foundry [6] and a C.A.D. vendor [7] who supply their manuals in HTML.
We can observe that many industrial documents are in "PDF", "Frame Maker" or "Interleaf" format. PDF material is very portable, but hardly support hyperlinks, while Frame Maker and Interleaf documents have little portability, requiring the use of very expensive editors and the installation of expensive viewers. In addition, these formats offer no possibility of fine level interactivity, while a HTML document containing embedded Javascript or Java is the best compromise between portability and interactivity.

An additional benefit of the choice of HTML is the possibility of making the courseware available on the institution's WWW server [8], enabling the students to access it from outside, to show it to their friends, family, etc...

3 Courseware organization

The courseware contains 4 types of document strongly linked together, as depicted on *figure 1*.
The lecture is not a complete course on analog circuits (this is not the goal of this action) but rather a selection of topics that can be illustrated efficiently during the practical training.
The exercises are the "core" of the courseware.
The "tools manuals" included in the courseware are digests of the user's manuals of the respective tools, with addition of some site-specific information (how to configure the environment, to edit text, to run the programs).

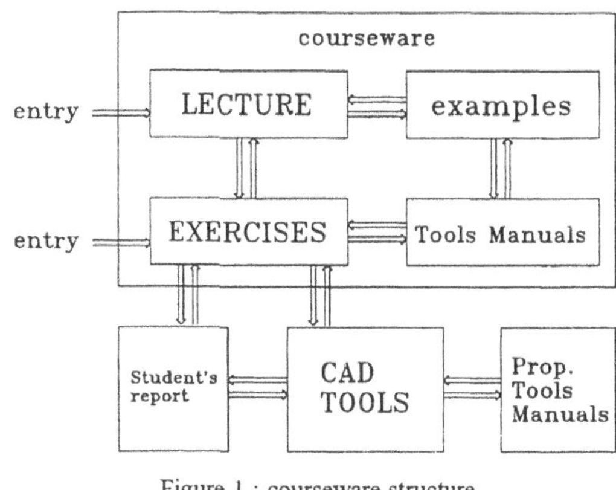

Figure 1 : courseware structure

Some of these manuals are written in French, some are intentionnally left in English.
There are two main entry points to the courseware : one to the lecture (if the subject is new for the user), one direct to the exercise, with possibility of consulting the lecture in case of doubt.
On *figure 1*, the arrows inside the courseware box represent actual hyperlinks, while outside they represent more abstract relationships.
Nevertheless, some tools (i.e. Cadence) have real hyperlinks between the tool and the documentation : this is known as "context-sensitive" help.
The student's report can include data generated by the C.A.D. tool, for example waveform plot pictures.
It is only for technical reasons that there is no hyperlink between the tools manuals summaries included in the courseware and the proprietary manuals of the same tools.

4 Analog circuits exercises

The main flow of exercises proposed in 1997-98 is the progressive construction of a CMOS operational amplifier, with intensive use of HSPICE for electrical parameters extaction and transistor sizing. *Table 1* gives a summary of the first steps of this sequence. Items marked with a * in the table are parameters to be inferred indirectly from simulation results, and are intended to recall that the basic laws of electricity must be constantly kept in mind.
This contruction is continued by the layout of the same circuit using Cadence (including relevant verifications : DRC, extract, LVS). Then an introduction of behavioral modeling of analog circuits is given, enforced with an exercise on the Sigma-Delta modulator.

5 Conclusion

This experience, carried with 42 students in 1997-98, appeared very rewarding. The students feedback was very positive, and they turned to be so enthusiastic that the amount and quality of the HTML reports that they produced exceeded the expectations. The teaching staff appreciated a lot the possibility of updating instantly the documentation, instead of having to print again and again a paper document. This gain in flexibility is considered as a compensation for the amount of preparation work. One problem arose with some documents which deserved to be also distributed on paper (e.g. SPICE fundamentals), because it is difficult to obtain a good paper document from HTML.

CIRCUIT	ACTION	EXPECTED RESULTS
given current mirror	DC analysis	output voltage range for a given current range, current range for a given output voltage range
	advanced DC analysis	input dynamic resistance, output dynamic resistance
improved current mirror	parameter sweep	MOS optimal sizing for improved output resistance and voltage range
differential stage using the optimized mirror	DC analysis	voltage gain with infinite impedance load, transconductance with zero impedance load, *output dynamic resistance
	AC analysis	cutoff frequency with infinite impedance load, cutoff frequency with current output on zero impedance load, *output node capacitance
same stage with inputs driven through series resistors	AC analysis	cutoff frequencies, *input capacitance, *Miller effect interpretation

Table 1 : summary of first exercises

6 References

[1] IEEE Transactions on Education, "Home Page", http://wmm.coe.ttu.edu/ieee_trans_ed

[2] Centre de Culture Scientifique Technique & Industrielle de Grenoble, "Silicapolis, l'univers de la microélectronique", *(CD-ROM, bilingual French-English)*

[3] A. del Rio, D. Valdes, "Three-Dimensional Model for Analog Circuit Instruction", IEEE Transactions on Education, November 1997, vol. 40, N° 4 *(HTML document)*

[4] A. Ferreira, J.L. Noullet, E. Sicard, "Analog Prof : A Computer Aided Teaching Software for Analog Circuits", proc. 3rd Mixed Design of Integrated Circuits and Systems Conference MIXDES'96, Lodz, Poland, June 1996, pp 628-631

[5] M. Chirico, F. Giudici, A. Sappia, A. M. Scapolla, "The Real Experiment eXecution Approach to Networking Courseware", IEEE Transactions on Education, November 1997, vol. 40, N° 4 *(HTML document)*

[6] Austria Mikro Systeme International, "AMS HIT-Kit Online Documentation", Version 3.00 - February 1997 *(HTML document)*

[7] Silvaco Data Systems, Inc. "The Virtual Wafer Fab Interactive Tools Tutorial" *(HTML document)*

[8] Atelier Interuniversitaire de Micro-Electronique de Toulouse (France), "Courseware Index", http://www.aime.insa-tlse.fr/cours *(HTML documents, mainly in French)*

A MULTIMEDIA EDUCATIVE TOOL CONCERNING MICROWAVE CIRCUITS, BASED ON A PHYSICAL APPROACH OF THEIR TIME-DOMAIN OPERATION.

M.R FRISCOURT, C. DALLE
IEMN UMR CNRS 9929, Avenue Poincaré BP 69
59652 Villeneuve d'Ascq Cedex, France
Tél : (33)3 20 19 79 25 Fax : (33)3 20 19 78 96
E-mail : cdalle@cam520.iemn.univ-lille1.fr

ABSTRACT

We are presently dealing with a multimedia educative tool in order to make easier the teaching of the complex physical phenomena governing microwave circuits and their related semiconductor devices. This course is systematically supported by results obtained from a general time domain electrical circuit model, the originality of which consists in the use of physical macroscopic semiconductor device models. The circuit model is used to generate time domain animations of both the circuit signal waveform and semiconductor device spatial internal physical quantities. A first demonstrator is devoted to microwave two-terminal transit time oscillators.

Two-terminal semiconductor devices are presently used in numerous microwave systems. They allow to realize the basic electronic functions such as RF power generation (IMPATT and GUNN diodes), signal control (Varactor diode), switching and RF power limitation (PIN diode)... The internal dynamic operation of most of these devices is governed by complex space charge effects leading to a highly non linear electrical behaviour. Moreover, they are mainly used under large signal conditions, transient or switching operating modes. Consequently, a clear, fast and accurate description of the operation of this kind of semiconductor devices, by means of simple concepts, such as analytical or electrical scheme based theoretical models is not obvious in the framework of a course. This feature can involve imperfect and superficial knowledge of students. On the other hand, a time domain electrical physical modelling allows to accurately investigate the dynamic properties of semiconductor devices. This kind of modelling is mainly developed and commonly used in research context.

However, it can also potentially constitute an efficient teaching tool when associated with modern multimedia means. In the framework of our research activities, we have developed a general time domain electrical circuit model relying on the nodal analysis. Its originality is to be based on the use of macroscopic physical semiconductor device models : one dimensional, quasi-bidimensional, full bidimensional drift-diffusion and hydrodynamic ones. Thus, this circuit model is well suited to the modelling of most of

T.J. Mouthaan and C. Salm (eds.), Microelectronics Education, 135-138.

the microwave circuits using two-terminal devices or transistors (MESFET, HEMT, HBT..). It allows to accurately describe the circuit electrical signal waveforms as well as to access to the space and time evolutions of the semiconductor device internal physical quantities. It is well suited in the simulation of transient and CW, linear and large signal operating modes over a wide frequency range up to the millimeter-wave one. Thus, we are presently developing an educational university project devoted to the physical approach of the microwave semiconductor device and associated circuit operation. This course is systematically illustrated by means of animations obtained from our time-domain circuit model making easy the understanding of the complex phenomena involved into the semiconductor device active zone.

A first course is devoted to the transit time semiconductor devices for RF power generation namely the IMPATT and GUNN diodes. It presents an arborescent structure constituted by a main course and additional facultative pages allowing the reader to more or less thoroughly study the different developed aspects. In the main course, the theoretical concepts related to the ideal transit-time negative resistance monochromatic oscillator modelling are progressively introduced. The influence of the passive load-circuit frequency behaviour on the global oscillator operation is investigated. This allows to define the ideal active device external signal waveforms leading to RF power generation and then to introduce the potential interest of transit time devices. The main physical phenomena governing their operation principle are described by means of a simplified device model. This model is used to investigate and define the required operating conditions and related device parameters leading to high efficiency RF power generation at a fixed frequency.

Then it is described how transit time operation at microwave frequencies can be obtained from semiconductor structures. The first one is the IMPATT diode. The RF operation of the single-drift Read structure is exhaustively analyzed. Its performance limiting effects are defined. CW and pulsed operating modes are considered. The second transit time device is the transferred-electron one, namely the GUNN diode. The course is focused on the analysis of the accumulation layer and transit time operating mode currently observed in millimeter-wave mesa structures.

The course has been written in HTML and is systematically illustrated by means of animated Gif sequences. A demonstrator is presently available on Internet at : http://www .univ-lille1.fr/~eudil/mrfcd/cours1.htm. In a near future, we intend to extend the course to the PN junction, the PIN diode and its related applications and later to transistor circuits. Our method is not limited to microwave circuits but could be extended to the study of any analog electronical circuit. We wish to develop an english version (but animations are comprehensive whatever your language is !). The figure is an example of an HTML page.

6 - GENERATION DE PUISSANCE PAR DIODE A AVALANCHE ET TEMPS DE TRANSIT

6.4 - FONCTIONNEMENT EN REGIME CONTINU D'OSCILLATIONS

➜ 6.4.2 - TRANSIT

Le plan de charge généré dans la zone d'avalanche pendant la première alternance du signal est injecté dans la zone de transit à l'instant t=T/2. Il y dérive à vitesse constante et égale à la vitesse limite dans le matériau (dite vitesse saturée).

Rappelons qu'il s'agit là d'une charge d'électrons, les trous ayant été rapidement absorbés par le contact P^+.

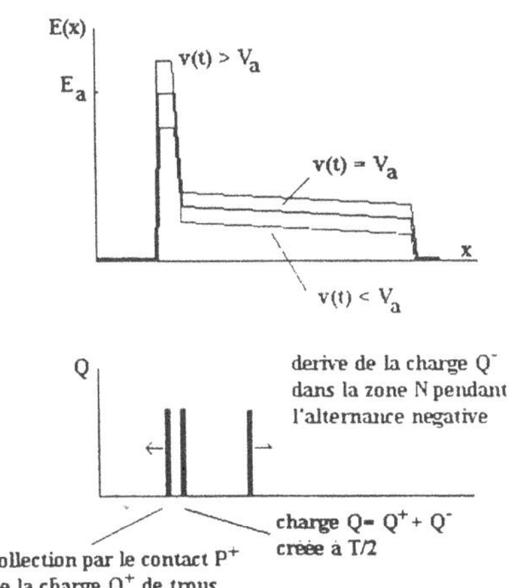

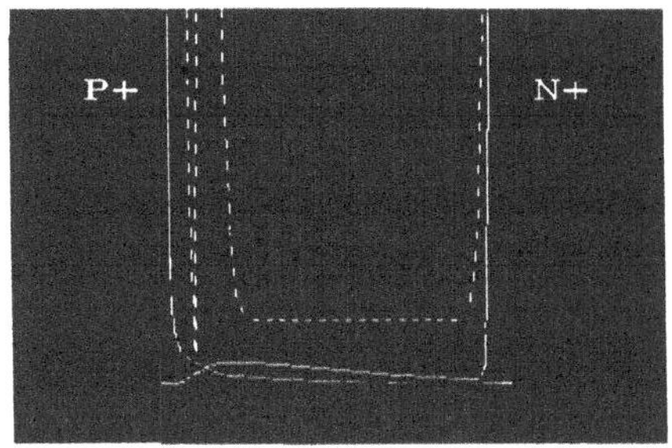

D'après le théorème de Ramo-Shockley, une charge d'amplitude constante se déplaçant à vitesse constante dans une région de longueur fixe, induit un courant constant dans le circuit extérieur.

Le temps de transit de la charge d'espace dépend linéairement de la longueur de la zone de transit :

$$\tau = \frac{L_t}{v_s}$$

Ainsi, si l'on choisit la longueur de la zone de transit telle que le temps de transit soit égal à une demi-période, le déphasage entre la tension et la composante fondamentale du courant sera de 180°, et l'on observera les évolutions temporelles suivantes :

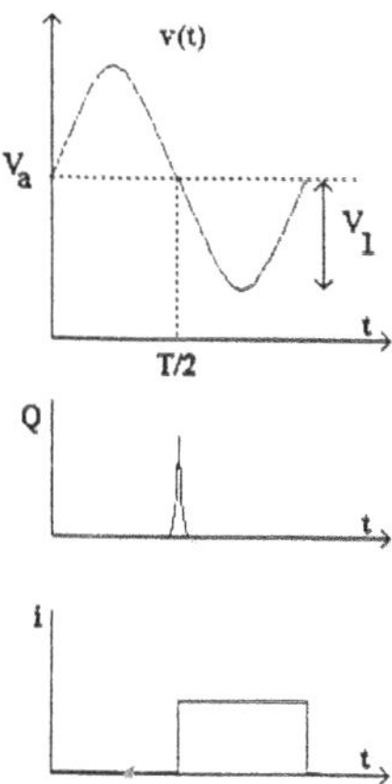

Le courant représenté ici est le courant induit dans le circuit extérieur par le courant de conduction. Afin d'obtenir l'allure du courant total dans le circuit extérieur, il nous faut maintenant tenir compte du courant de déplacement.

DISTANCE EDUCATION IN MICROELECTRONICS VIA MULTIMEDIA

NICULAE RADU, ADRIAN DRONDOE, EMIL RĂDULESCU, GABRIEL DIMA, CHRIS VON KOSCHEMBAHR*, MARCEL PROFIRESCU
EDIL R&D Centre, Technical University of Bucharest
PO Box 57-112, Bucharest 74500, Romania
**IBM Global Services, Education & Training*
Chaussée de Bruxelles 135, B-1310 La Hulpe, Belgium

1 Abstract

A Distance Education in Microelectronics Network via Multimedia is presented. The IBM Personal Learning System Software used as course administration environment offers a unique combination of interactive multimedia distance learning, open courseware development, distribution and management and student tracking.

2 The Need for Distance Learning

The present development of information and communication technologies is a real stimulus for universities, which are no doubt challenged to redefine their strategies and functions. Secondly, the growing importance of open and distance learning in the "digital era" will determine a shift of present teaching methods. In the past, the computers were largely used for huge and intricate computations but now they also became tools integrated in the teaching process. Computer Assisted Teaching (CAT) and Computer Based Training (CBT) are alternatives to classical teaching. This concept offers many opportunities for youngster and adults to enroll in the program that meets their interests and needs, at the convenient time and duration, in a learning centre or on their laptop computer.

EDIL R&D Centre is engaged in developing the infrastructure and the content for the first academic distance learning network in Romania. The project is supported by the National Higher Education Funding Council for three years and at the end of this period the project is going to support itself through contracts with different customers.

The project started almost one year ago and now it delivers distance education lectures in Microelectronics for students from Faculty of Electronics and Telecommunications in the Technical University of Bucharest. In the frame of this project our Centre is associated with three other universities in Romania forming a distance learning academic network in which all these universities share their courses. The third party in this project is IBM Global Services Education & Training in La Hulpe which provides the technical support and software for the distance learning network.

T.J. Mouthaan and C. Salm (eds.), Microelectronics Education, 139-142.

3 Hardware Implementation

Each university has a PC LAN dedicated to distance learning that is connected via Internet with partners' networks. The IBM Personal Learning System (PLS) runs on the local server at which other multimedia PCs have access. PLS is optimized to use minimum hardware resources such that it runs also on standard multimedia PCs that offer the students interactivity, sound and synchronized image motion. These PCs are relatively inexpensive so that anyone could afford to have one and take lectures right from their home through a modem connection.

The network configuration is presented in Figure 1.

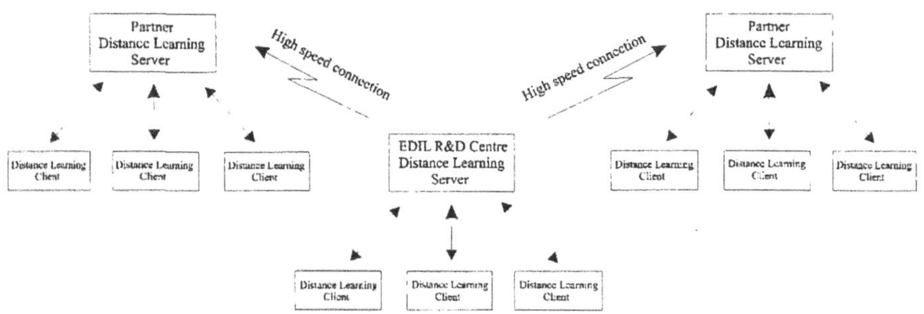

Figure 1: The PLS Distance Learning Computer Network

4 Personal Learning System Software

PLS is a software package that manages and delivers courses, assures the user's logging and monitors the performance of the students by sending evaluation tests to the users. The course administrator receives the answers and assures the interaction with the students through written messages (mail).

PLS provides access from a personal computer to the required content. The content may be traditional computer based training, digital video (Figure 2), cable or satellite TV or digital audio. The main advantage of PLS is the possibility to present multimedia courseware and tests. These could be already made files in a standard format (html, doc, avi, mpeg, ppt) or teacher created programs (executables). The color images, animations, digital sound, TV facilities and interactive programs keep the student attention alive and challenge him to overpass the obstacles met in the classical teaching process.

PLS for Windows is a software tool that offers a unique combination of interactive multimedia distance learning on demand, open courseware development and delivery platform, quick and easy multimedia authoring, networked audio and video distribution, multi-lingual support, open courseware distribution and management, student tracking. It also provides support for database (dBase IV, DB2/2).

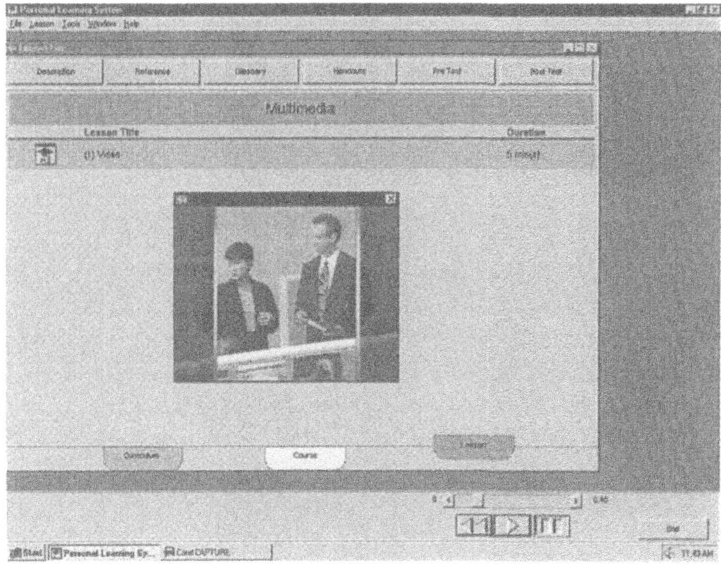

Figure 2: A Multimedia Application

Figure 3 presents the PLS interface and an application included in the IC Simulation and Design course. The carriers movement and the generation/recombination processes in an abrupt pn junction could be better understood from animated pictures than from an usual picture from a book. The course also includes multimedia applications that simulates the technological processes involved in the IC manufacturing and the electrical behavior of semiconductor devices.

5 Impact on the Target People

After the PLS implementation in EDIL the student average marks increased showing that the teaching process is more efficient and students have gained increased knowledge and skills. A PLS feature allows the students to rewind the lessons as many times as they need, in order to review the weaker points and have access to instructor's expertise.

One could consider that in short run this teaching method is more expensive but doubtless it is more efficient and it is always open for updates to the latest reported results in microelectronics.

6 Future Directions in Developing the Distance Learning Network

The PLS version 2.3 is designed to be used in a local network. The users ask for a course located on the server or on other computers in the LAN. The course exchange takes place using FTP over the Intranet network or are exported on CDs to the customers. This version is appropriate for large files and high speed execution.

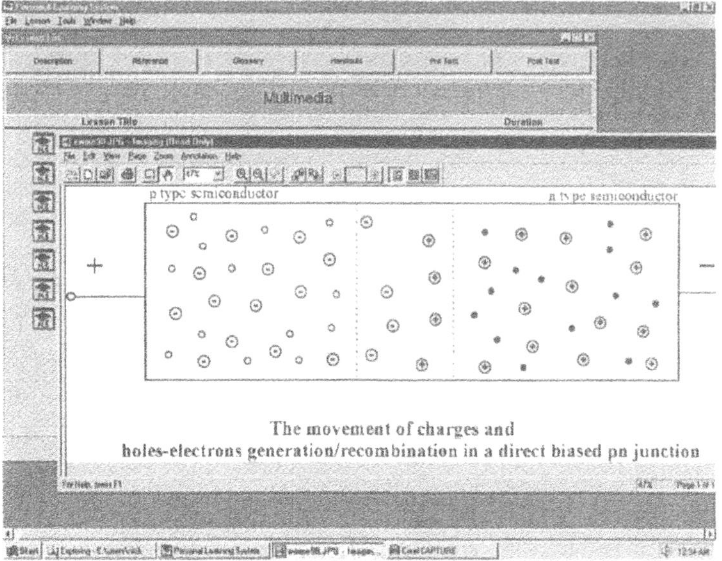

Figure 3: ThePLS Interface of the Course on IC Simulation and Design

The PLS Notes version allows access to the same features, previously described, now from a Web Browser and Lotus Notes Clients. It adds new possibilities for students to collaborate with other students and instructors.

The described continuous education network is a pilot for a national network for distance and continuous learning which is going to improve and make the teaching process in universities more efficient. This method is targeted to undergraduate and graduate students, but also specialists or people that desire to enlarge their horizon of knowledge.

7 References

1. IBM Personal Learning System, version 2.3 manuals, 1997.
2. IEEE MultiMedia - vol. Jan.-Dec. 1998.
3. M.D. Profirescu, G. Dima, Integrated Circuits Simulation and Design, Technical University of Bucharest Press, 1995.

AN INTERACTIVE MULTIMEDIA COURSE ON BASIC DIGITAL ELECTRONICS

D. DEL CORSO*, P. CIVERA*, V. POZZOLO*,
C. SANSOE'*, C. SCRIZZI^
Politecnico di Torino, Department of Electronics (*) and Multimedia Laboratory of Politecnico and COREP(^)
C.so Duca degli Abruzzi 24, 10129 - Torino - ITALY

1 Introduction

This work describes the organization and the educational structure of an interactive multimedia course on Basic Digital Electronics, with a short presentation of the development environment (designed to simplify the task of content authors), and of the user interface (designed to maximize the educational effectiveness).

The development of educational interactive multimedia packages, even for well settled subjects, requires careful didactic design and tools to ease the transfer of contents from existing media (books, notes, ...) to the new ones. Our approach for the development of an educational package starts with the definition of a *structure*, that is a hierarchical sequence of *pages*, freely planned by the author following the directives of a *didactic guide*. An interactive *structure generator* builds the framework while a *rule book* guides the author providing information on compatibility constraints, and page prototypes. The structure includes a *user interface* with an ample choice of navigation tools and other commands. The environments drives the author towards correctness by constructions; however, a *test suite* can be applied to final products, to verify both the functionality and the effectiveness from the educational point of view.

With interactive educational packages, the "doing" dimension is added to listening and seeing; that means the freedom to choose place, time, speed, and path for learning, and direct interaction with the tool itself [1]. In every page of this package the student is forced to perform some actions (from buttonclick to using simulators or answering simple questions) to explore the content and proceed to following subjects.

2 The Educational Approach

Since the beginning the work aimed to the production of complementary packages, which do not replace standard lessons but can be used by teachers to enhance room les-

T.J. Mouthaan and C. Salm (eds.), Microelectronics Education, 143-146.

sons, and by students to review theory, practise exercises, and perform interactive simulations in a "virtual laboratory". This package is addressed mainly to courses in Applied Electronics for Computer Engineering, and assumes preliminary knowledge on Analog Electronics. The content is organized in four chapters:

1. Switching Transistors;
2. Logic Circuits;
3. Signal generators;
4. Interconnections and interfaces.

The last Chapter addresses subjects such as fundamentals of transmission lines and signal integrity analysis, now mandatory for high speed digital circuits designers. Contents are co-ordinated with videotape lessons (from the NETTUNO consortium).

The work is organized in a hierarchy of Chapters/Lessons/Subjects. For each subject, the content presentation follows the typical sequence for technical matters in engineering courses:

- qualitative description;
- quantitative analysis and development of a mathematical model;
- application of the model to some examples;
- direct interaction with the model using interactive simulators (virtual laboratory);
- experimental verification;
- tests and numerical exercise.

Multimedia technology is best exploited in the last steps: The interactive simulator (4) allows the student to carry out "virtual experiments". Interaction and presentation of results occur through a graphic user interface. Step 5 is a laboratory experience, with a step-by step guide and a short movie to show execution and results. Tests and exercises span from multiple choice questions, to step-by-step problem solving, with links to "theory" pages in case of mistakes.

A "teacher edition" of the package (currently under development) embeds a basic authoring system, which allows users to add text, graphics, audio and video clips to any page. This tool enables each teacher to build personalised and enhanced versions of the package.

3 The Q-LAMP Environment

Quick-LAMP (Q-LAMP) is a set of tools developed at the Multimedia Laboratory of the Politecnico di Torino and COREP (LAMP) to support quick prototyping of educational interactive multimedia [2]. It is based on a commercial authoring environment (Toolbook from Asymetrix), but the same methodology can be applied to similar packages.

The tools available in Q-LAMP include:

- *didactic guide*: with examples of organisation for educational packages;
- *rule book*: defines rules to concatenate contribution from different authors;
- *structure generator*: builds content books organised in a hierarchy of chapters/lessons/subjects, and allows to modify their organisation.
- *test and exercise editor*: to build exercises and keep track of user performance;
- *testing suite*: a checklist for the final products;
- *assessment tools*: to track student behaviour for product effectiveness eveluation.

The above described tools make available to the author:

- a set of *content books*: built by the structure generator, they embed the actual technical information to be delivered to the user;
- a *system book*: handles navigation and other functions for all content books.
- a library of *scripts and graphics* for interactive animation.

Animation and interactive graphics are extensively used. 2-D animation use the Open Script language (Toolbook native language), while 3-D animation and movies are prepared using external tools. Simple interactive simulators use Open Script code, thus keeping real-time interaction. For complex arithmetic functions an external simulator (e.g. Mathlab or Pspice) is called from the Toolbook page.

4 The User Interface

The package is organised like a textbook, with Chapters, Lessons, Subjects, but the hypertext links allow the user to follow different paths in the structure. The smallest didactic unit is the *Subject* (one or a few pages); they are grouped in *Lessons,* designed for sequential exploration, but the user can directly reach each Subject from the index pages or using the navigation tools.

In each page an *information bar* helps the reader to identify current position (Chapter, Lesson, page number, and progress within the lesson), and a *command bar* makes available a set of commands for navigation and other functions.

Frequently used commands, directly available from the command bar are:

- global navigation tools (Map, Table-of-contents);
- notebook and history;
- local navigation commands: move to adjacent pages, back to previous page;
- "emergency" buttons: help, exit.

Navigation tools, besides the usual next/previous page commands, include:

- *History*: a list of visited pages, built by the systems during navigation.
- *Index*: a table of contents; the user can move from it to any page in any chapter.
- *Map*: a graphical list of chapters, lessons, and subjects organised in a tree structure.

Index and History have text *find* and *go-to* capability. History can be used to define specific sequences of pages (to prepare sessions of guided navigation). The notebook accepts text notes associated with each page. User-specific history and notes can be saved and re-loaded. Other seldom used commands are available with two-step operations.

The "teacher" package has additional commands which allows the users to add text, graphics, audio and video clips to any page. These new items can be saved and copied to students' machines.

5. Conclusions and Perspectives

The experience developed with the course on Basic Digital Electronics has been exploited also for packages on Electromagnetic Compatibility [3], Safety in Industries [4], and several other subjects. The package on Digital Electronics is now used in courses at the university and is commercially available for external users.

The Q-LAMP environment allowed to cut development time, especially for complex and articulated educational packages. As an example, the complete navigation structure for a package of about 2000 pages has been developed in a day. The time saving and the fact that the content author is relieved from technical issues related to authoring tools make available more resources for the actual didactic planning, thus increasing the quality and effectiveness of the final product.

Current activities, besides the development of new contents, address mixed distribution technologies (CD + Internet), quality evaluation tools, assessment of the educational effectiveness, and improvements in the structure generator with the goal to ease the preparation of multi-language packages and give more freedom to the authors and users for the development of "custom" modules.

Acknowledgements

The work has been carried out at the Multimedia Laboratory (LAMP) of the Politecnico di Torino and COREP. The Basic Digital Electronics module and the Quick-LAMP environment have been developed and tested in several graduation thesis from 1995 to 1997; besides the authors of this paper, the list of people involved includes F. Dell'Apa, D. Di Polito, M. Gastaldi, S. Imarisio, E. Ovcin, L. Placentino. The work has been supported by COREP and by the Politecnico di Torino.

References

[1] A. Scarafiotti et al.: An Experiment in Interactive MultiMedia Learning Tools, 8th Conf. on College Teaching and Learning, Jacksonville, FL, April 1997.
[2] D. Del Corso et al,: Q-LAMP: an Environment for Quick Development of Educational Interactive Multimedia Packages, EMMSEC97, Florence, November 1997.
[3] F. Canavero et al: Electromagnetic Compatibility - ECCE Project, EU Leonardo Program, 1997.
[4] D. Vidotto et al: Safety in Industries - SICURO Project, COREP and Ministry of Work, Italy, 1997.

OPTIMISATION OF A SOLAR CELL

Exercise in simulation based training using the MODEM Tele-learning concept

Jan Hovius*, Henny Leemkuil*, Jan Hensgens, Ton Mouthaan*, Cora Salm***
**University of Twente, The Netherlands,*
***RIKS Maastricht, The Netherlands*

1. Abstract

Simulation based education supports active learning, enables students to acquire knowledge easier and get acquainted with the professional simulation software.
Using Telematics within the courses introduces more openness and flexibility: the courses can be kept more up-to-date and adjustments to specific user needs can be implemented easier. Next to this it creates new training facilities at a distance, e.g. Tele-teaching, Tele-conferencing or Tele-assistance. MODEM – Multimedia Optimization and Demonstration of Education in Microelectronics – an EC funded project exploits both technologies to establish a European wide training service for the microelectronics industry and higher education. MODEM uses a virtual campus by which access is given to its services. A one week course – Optimization of a Solar Cell – has been developed to demonstrate the concepts of using Telematics together with simulation based exercises. This paper discusses the course itself and the evaluation results of a pilot run.

2. Introduction

The MODEM project is focussed on exploiting new and emerging techniques in the area of simulation based learning, co-authoring and flexible delivery in order to enable better use and access to knowledge and resources available in the field of microelectronics. The goal of the project is to investigate the feasibility of a self-sustaining European wide training service for the microelectronics industry and higher education. This will be done using a *virtual campus,* a sort of market place where needs for education and training in microelectronics can be met with tailor made modules, courses and exercises. Amongst others, Tele-based courses are developed enabling peer learning as well as group exercises using synchronous and asynchronous techniques to communicate. One of the exercises, developed to show (some of) the concepts of this project, is a Tele-based interactive course on the optimization of solar cells. This course covers several Tele-techniques like: Stand-alone teaching with real-time access to a remote simulation server, peer learning, in which a group of students need to collaborate to solve a task, using shared resources (shared whiteboard or server application) and Tele-assistance using either synchronous (Chatbox) or asynchronous (E-mail) tools.

T.J. Mouthaan and C. Salm (eds.), Microelectronics Education, 147-150.

3. The Solar Cell Course

The solar cell course is derived from an already operational laboratory assignment given at the University of Twente as part of a broader education project ("Trimester 3.1 Integration project"). The text-based instruction handouts and always-available tutor have been replaced with MODEM's simulation based learning environment (Fig 1).

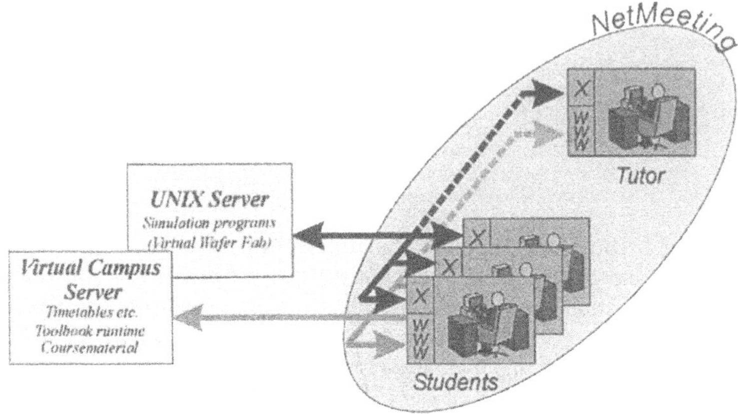

Figure 1: The MODEM learning environment.

Figure 2 shows a screenshot of the solar cell course running the Virtual Wafer Fab (left), the Toolbook course itself (upper right) and a Chatbox session (NetMeeting).

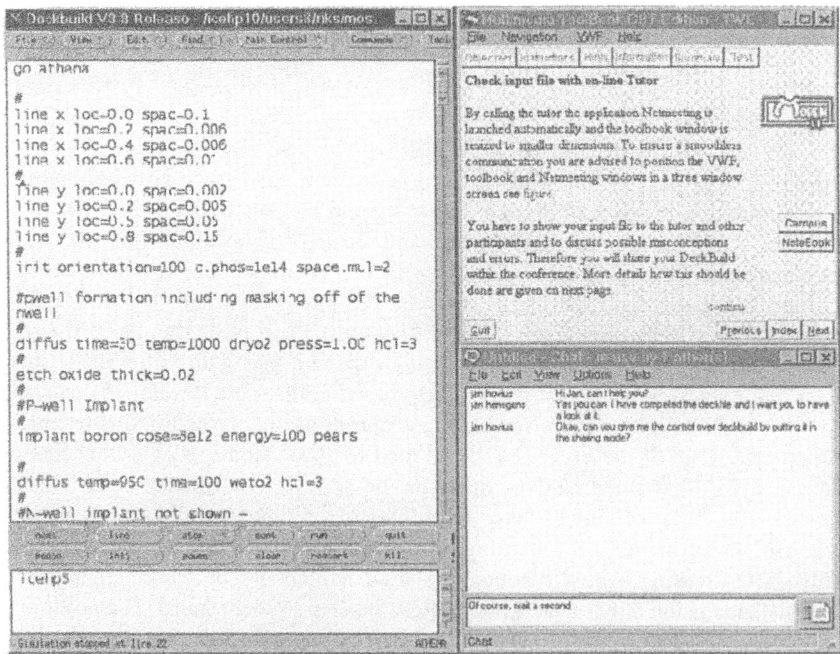

Figure 2: Screenshot of the Solar Cell Course: Left a simulation window (VWF's Deckbuild), the course itself in the upper right corner and a Chatbox (NetMeeting) in the lower right corner.

Since the course is designed with Asymetrix's Multimedia Toolbook©, every student needs to have a runtime Toolbook library installed on his workstation (downloadable from the virtual campus) as a pre-requisite. Next to this Microsoft's NetMeeting© (shipped with Internet Explorer 4.0) is used for communication purposes and finally an X-Server package (Hcl Exceed©) takes care for communicating with the X-based Simulation package (Silvaco's Virtual Wafer Fab©) running centrally on a UNIX server. The exercise focuses on optimizing a solar cell device for maximum power efficiency when used in sun light conditions. By varying junction depth, doping level of the top layer, properties of the anti-reflection coating and layout of the metallisation of the top electrode the maximum achievable power output has to be determined. Device simulation using Silvaco's *Virtual Wafer Fab* and circuit simulation using SPICE provide the means for determining the electrical output of the device. The exercise itself is preceded by a short theoretical study on the operation principle of a solar cell together with a short hands-on exercise exploring (some of) VWF's capabilities. The students conclude their assignment with a final report on the results obtained and send this via E-mail to the tutor.

The reserved time for completing this course has been determined at 40 hours. This is a significant longer period than is involved in working with "traditional" courseware.

4. Pedagogic scenario of the course

The pedagogic scenario of The Solar Cell course has clear constructivistic aspects, uses the Internet environment and has incorporated aspects of collaborative learning. Students are expected to be actively constructing knowledge by seeking for information and testing hypotheses, and discussing results with others. To give them some sort of guidance in this complicated environment and complex task, some form of scaffolding is implemented and some other elements to reduce cognitive load:

- To make sure the students are active and can do simulations quickly VWF commands are presented combined with copy/paste facilities.
- Simulations can be started by pressing buttons in the courseware program
- Information is available about the course structure and objectives
- Background information about VWF is available
- Important figures and tables can be enlarged, and are also available via the menubar
- Each sub-task contains some open questions and assignments
- To get help while solving questions/tasks hints are available
- Students are promoted by the instructional program to discuss answers or work together with others and can contact other students or a tutor by pressing certain buttons (starting NetMeeting or E-mail facilities)
- There is a link to an editor to make notes and write down answers to questions and assignments.

5. Pilot run, evaluation and conclusions

A small group of students from the department of Electrical Engineering of the University of Twente evaluated the course using pre-configured workstations (WINTEL based) connected to the Internet via the TCP/IP protocol. They were working at different locations. All actions undertaken were recorded by observers and evaluated after each session.

An overview of the results, as extracted from the filled in questionnaires (based on Raven & Johnson (1989)) is given.

- The students gave the course a high score on aspects like learnability, functionality, navigability, interface design, speed, workload and precision.
- A less good score was obtained on aspects like helpfulness, controlability, likeability and groupwork suitability.

Some of the remarks sorted by scenario used:

- *Stand alone* – The students didn't encounter many problems using the course in a stand alone setting except for becoming familiar with the VWF environment (different Windows/Mouse definitions).
- *Tele assistance* – Main problem encountered with this discipline was a good planning of the (pre defined) time slots at which the tutor was available for online assistance: since every student works at his/her own pace it appeared to be difficult to schedule such synchronized sessions. This sometimes forced students to stop the session and wait! It was remarkable that the use of a trial and error strategy was preferred over consulting the online manuals in such a case.
- *Group work* – In this setting a group of students need to solve a problem together using the Chatbox and shared applications. Except for some small problems (bugs) with the communication software (NetMeeting) chatting and sharing applications didn't give rise to complaints from a technical point of view. On the other hand it appeared to be difficult to use such tools with more than two people involved. You need to be very disciplinary in order to prevent "mouse wars" (who's in control of the application) and chaos in the Chatbox.

As a result of the pilot run some general conclusions towards Telematics based courses and the Solar Cell course in particular, can be drawn:

- The acceptance of online communication tools (E-mail, Chatbox, and Whiteboard) doesn't appear to be problematic. In this era of internetting those tools are already of common use.
- The Internet can provide an efficient way for online sharing of expensive resources (e.g., remote use of professional TCAD software).

6. References

Hensgens,J., Rosmalen, P. van, Hovius, J., Mouthaan, T., Merriënboer, J. van & Leemkuil, H. (1997). ***Network-based training for microelectronics using simulation tools.*** Paper presented at the European Materials Research Society Spring Meeting in Strassbourg (June 16-20, 1997).

Leemkuil, H., Meikle, F. & Trayhurn, D. (1998). ***MODEM Evaluation report.*** Deliverable 7.2[1]

[1] MODEM deliverables can be downloaded from the following WWW-location: http://nmrc.ucc.ie/projects/modem/deliver.html

VLSI DESIGN IN THE MODEM PROJECT

U.S. LIDHOLM, A. CHAGOYA, R. LEVEUGLE
NMRC *CIME-INPG*
Cork, Ireland *Grenoble, France*

1 Introduction

This paper describes a production activity of the MODEM VLSI subject matter experts within the MODEM Production Environment.
The objective was to produce a multimedia learning material demonstrator to provide both a case study for the MODEM toolbox and also to act as demonstrator to the wider industrial and academic microelectronics community. In particular methods of co-authoring have been explored during the development of the demonstrator.

2 The Demonstrator

The VLSI Design Demonstrator was designed according to the three main objectives
- to take advantage of the computer environment to go beyond classical course material
- to promote more initiative from students ("active" learning rather than just listening)
- to make it possible for each student to adjust the time spent on each part of the course

The contents of the demonstrator is divided into two independent subject groups:
(1) Memory Cells: Dynamic Memory, Static Memory and Semi-Static Memory
(2) Registers: Master Slave Registers and Gate Based Registers
The general script layout consists of an input menu group addressing subject modules within each subject group. The subject modules include a tutorial part which presents the design procedures and knowledge the student has to understand and an "exercise" part for evaluation purposes. Each subject module is structured in a set of sections preceded by a general introduction and culminated by the general conclusions. Specific introduction and conclusion for every section are also provided. Some sections include a combination of animation, simulation and an exercise depending on the nature of the topic being covered.
When the user has gone through the tutorial part, labelled navigation tools become visible with explanatory text, allowing either reviewing the tutorial, visiting the animation or exercise, or entering into the simulation part. See figure 1.
It is assumed that the learner is familiar with the basic concepts of microelectronics. More specifically, basic knowledge about logic gates, the basic behaviour of a MOS transistors and the understanding of the hierarchical description of abstraction levels in terms of logic and transistor circuits.

T.J. Mouthaan and C. Salm (eds.), Microelectronics Education, 151-154.

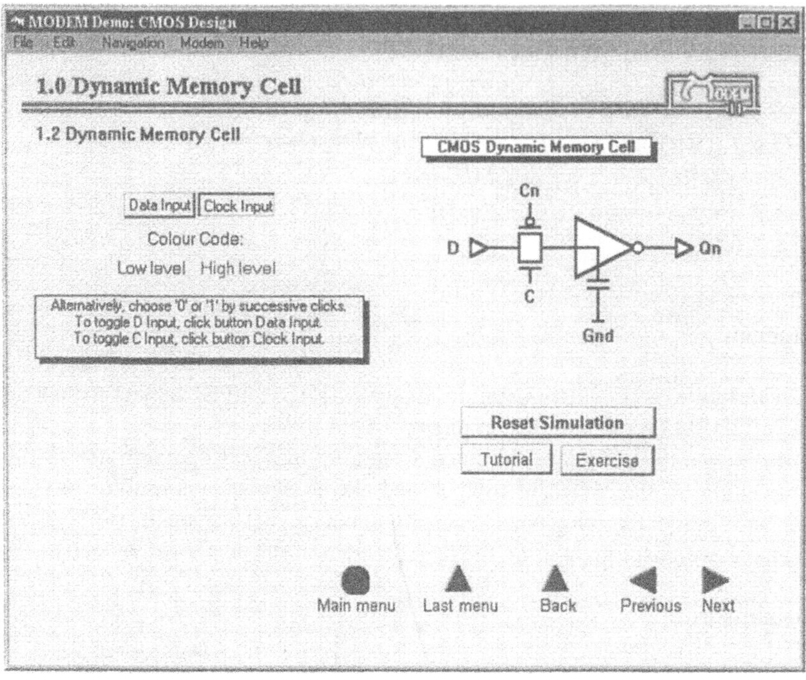

Figure 1. Example of navigation tools allowing either reviewing the tutorial, visiting the exercise, or entering into the simulation part.

3 Interaction

The course is designed in such a way that the student gets actively involved in the learning process as opposed to the classical methods. The basic interactions ensure activity by the student, in particular the sequence of screen control, e.g.:

- clicking on the mouse once the concept is understood to go to the next step
- having the possibility to go back to relevant knowledge (or to display additional information) by means of hypertext and viewers
- answering questions or giving suggestions (with feedback from the multimedia environment) instead of just reading affirmations.

To avoid similarities with a classical book, course pages are not immediately filled upon entrance but instead, in a step by step manner under the user's control via button clicks.

Typically, text information is located on the left hand side of the screen and electronic schematics on the right hand side. Text is also displayed in the schematics with suitable highlighting. Additional information concerning the elements of the schematics is provided by mouse operations on corresponding symbols.

3.1 ANIMATION

Animation sequences are intended to show structural evolution and thus broadening the explanations supplied. Changing the colour of part of a schematic, or flashing a particular element on the screen is carried out to clearly identify an element relevant to an explanation. Operation of electronic circuits can be shown in a very attractive way using animated sequences, with colour changes representing the way a signal propagates between circuit nodes.

3.2 SIMULATION

Simulation sequences provide a deeper insight into the topic just presented. Selected simulation sequences of propagating 1's and 0's are introduced in the demonstrator for the sake of a better understanding of the functional aspects of the design.

3.3 EXERCISES

Exercises are not used for evaluation purposes but instead to help the user to detect potential misunderstandings. The exact type of exercise depends on the course contents but also on the features of the course environment. Ideally, the students should be able to actually interact, and not only to choose an answer. Therefore not only multiple choice questions are used, but also exercises involving schematic-capture.
A logic symbol browser can be invoked and used with schematic-capture oriented exercises. A waste basket is provided on pages where schematic capture is used to get rid of symbols unnecessarily picked up from the browser. See Figure 2.
A glossary is provided as a separate book, strongly linked to the main book. It includes textual explanations and a summary list with all concepts documented. The summary list is implemented using hypertext to allow easy navigation inside the glossary. The glossary can be explicitly invoked from the navigation window or indirectly, by mouse operation on the symbols of the different schematics presented in the course.

4 Co-authoring

Co-authoring has been successfully put in practice in three main developing activities: the implementation of the Modem Style Guide recommendations, the modification of electronics schematics and the revision of terminology and textual descriptions.
NetMeeting sessions started at a time fixed before hand. Parts of the discussions have been held through the Internet using the sound capabilities of the NetMeeting software but it has not been always possible because the shared application unfortunately introduced a lot of noise on the sound. In this case the telephone has been used as an alternative. In the cases where a colleague from ISA (Denmark) or from PHASE (France) joined the co-authoring session, the chat box was essentially used for our joint discussions.

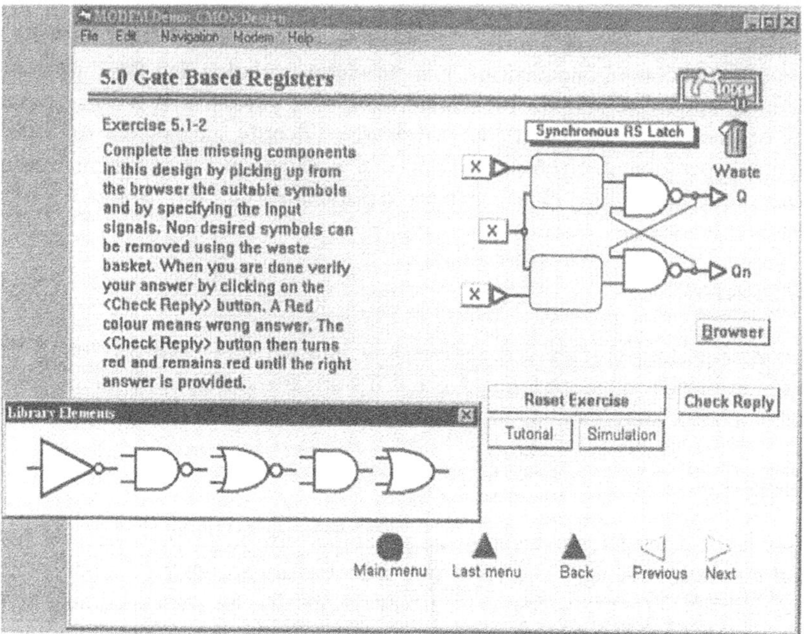

Figure 2. The browser is a pallet of basic logic symbols intended for use with the schematic-capture oriented exercises.

5 Software

The software used was Multimedia ToolBook (V. 4.0 CBT edition from Asymetrix) as an authoring system, Microsoft NetMeeting (V. 1.0) as a co-authoring facility and the MODEM toolbox Comas (**Co**-authoring and **Ma**terial **S**election from Courseware Scandinavia) database environment run on Lotus Notes (Client 4.1) used for archiving courses and for pick-and-mix distribution within the MODEM environment.

6 Conclusion

The paper has presented multimedia material developed by subject experts within the MODEM project. The co-authoring experience gained by the participants during the project has been very positive and is worth to be continued.
The software tools used have proved to work very well.

7 Acknowledgement

This work was part of the MODEM project funded by the European Union Telematics Application Programme within EU Directorate Generale DGXIII.

MATERIALS ANALYSIS IN THE MODEM PROJECT

F. Pêcheux[1], Y. Hervé[1], H. Marchal[1],
N. Hertel[2], J.P. Stoquert[1], R. Stuck[1], P. Siffert[1],
CNRS/PHASE[1], BP 20, 67037 Strasbourg Cedex 2, France,
ISA[2], University of Aarhus, Ny Munkegade, DK-8000 Aarhus C, Denmark

Abstract

The paper presents a set of multimedia courses on material characterization in microelectronics. So far, XPS/ESCA and RBS have been implemented. The courses are the result of a common work between ISA and PHASE, and are validated through the European Project MODEM, « Multimedia Optimisation and Demonstration for Education in Microelectronics ». A special entity named « the virtual campus » in the MODEM terminology will at term integrate these courses and, thanks to Internet, make them available for the whole European Community, for both academic (postgraduate) and industrial purposes.

One of the major interest of the courses is the protocol simulator available for some of the characterization methods. Such a simulator aims at giving the user a good look and feel of the real and complex experimental setup by allowing him to perform actions on its simulated model, according to the real protocol of use. Thanks to an appropriate design methodology, it is pretty easy to adapt the protocol simulator engine to any industrial or academic specific setup and to couple it with other simulators (like spectrum simulator), thus providing virtual instrumentation on surface analysis.

1. Context

The courses are the results of co-authoring between ISA and PHASE through the European Project MODEM, "Multimedia Optimization and Demonstration for Education in Microelectronics". MODEM aims at exploiting new and emerging technologies in the areas of simulation assisted learning, co-authoring and flexible delivery by enhancing the contribution that the higher education sector can make in addressing the skills shortage in the European microelectronics industry. A special entity, called the "virtual campus" in the MODEM terminology will at term integrate these courses and, thanks to Internet and ISDN lines, make them available for the microelectronic community, for both academic (postgraduate) and industrial purposes. To enhance our pedagogical standards, the following multimedia material has been designed keeping in mind that education by simulation and learning by doing are crucial issues. The courses have been developed with the multimedia authoring software Asymetrix ToolBook 4.0 and are available on PCs running Windows 95 or via Internet. The contents has been specified by experts, the ergonomy and continuity have been done under the responsability of teachers, and the coding has been performed by several computer engineers.

2. Contents

From the student perspective, the principle beyond surface analysis is slightly the same for all methods, i.e. projecting incident particles of a given mass and energy on the sample to be analyzed and evaluating the energy, type and mass of created or scattered particles from the sample. For this reason, the courses all follow a common template composed of six parts : History, Overview, Theory, Instrumentation, Evaluation, and Protocol Simulator. The first five parts are an attempt to produce a state-of-the-art multimedia course on surface analysis. The latter part allows the user to interact with a virtual implementation of the characterization method set-up. So far, two courses on characterization have been achieved, XPS/ESCA [1], and RBS [3]. Where applicable, the courses introduce sensitive concepts dynamically, like kinematic factor, backscattering factor and Rutherford cross-section, by making an extensive use of three dimensional animations, sounds, and interactive graphics.

T.J. Mouthaan and C. Salm (eds.), Microelectronics Education, 155-158.

2.1 The five « classical » views

History: For postgraduate students, it is important to consider the historical aspects of the methods : when it has been discovered, by who, and under what conditions.
Overview: aims at giving the user a coarse view of the method. The basic principle of each method is presented, along with its characteristic results and performance. Each multimedia overview lasts about 5 minutes, when playing the corresponding animations and listening to the comments.

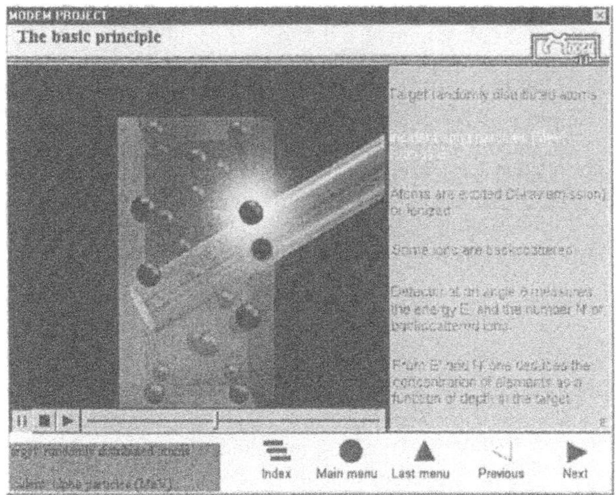

Figure 1 : The basic principle of the RBS method.

One can notice that the screen contains 3 main parts : the animation window where the basic principle is displayed and its associated VCR controls, the explanation window that reminds the key ideas, and the navigation window.
Theory: presents the underlying physical principle and formulae. It is by far the most important part for the student. From a pedagogical viewpoint, we have decided not to face the student with a linear list of items to be read, but on the contrary, to let him browse an hierarchical tree of fundamental notions which appear progressively, as he reads items. At the toplevel, 4 items are available, as shown in figure 2. Each time the user clicks on an item, the corresponding page is displayed.

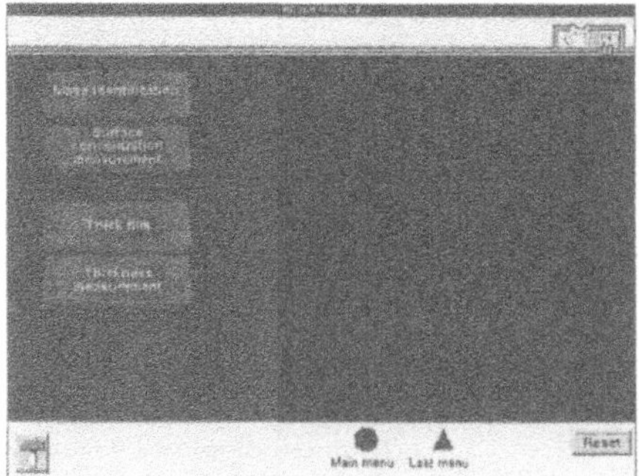

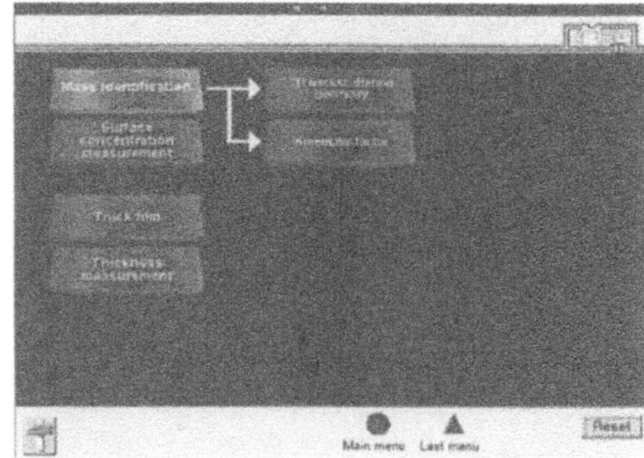

Figures 2 and 3 : The concept map of the « Theory » part, and the expandable tree of items.

When all the items of given level of knowledge have been read, the next time the user goes to the « Theory » concept map page, he gains access to more detailed informations, i.e. the corresponding terminal leaf of the tree is expanded into a node, which in turn contains new

items. Thanks to this approach, the student progresses at his own pace and is not overhelmed by too much information. Figure 3 shows how a terminal leaf is transformed into the corresponding node in the concept map tree. Some pages, as shown on figure, contain exercices that are direct applications of important notions. To help the user finding the solution, several specific tools have been developed on top of the multimedia courses, that can assist him on request. For instance, in the RBS course, the tool icon in the lower left corner gives access the periodic table of elements, to a kinematic factor table, to a Rutherford cross section table, and to a pocket calculator.
Instrumentation: displays a three dimensional representation of the setup to describe the basic components (accelerator, detector, target chamber, analysis devices,...).
Evaluation: faces the user with a suite of problems, from simple MCQs to complex material analysis problems. The latter exercices consist in finding the exact atomic composition of a material, starting with nothing but its RBS or XPS result spectrum. The same add-on tools (table of elements browser, stopping power table and pocket calculator) assist the user on request.

2.2 The protocol simulator

The last part of each course is entirely dedicated to the protocol simulator. Such a simulator aims at giving the user a good look and feel of the real experimental setup, so he quickly gets familiar with the manipulation of the different devices involved in a surface analysis experiment. It is therefore a practical counterpart to the previous theoretical sections of the course. In pratice, the user is faced with a graphical idealized representation of the setup and has to perform actions step by step according to the protocol of use of the real setup by clicking on a control panel or directly on the devices. Reading the status of devices is performed in a similar way.

The protocol is the suite of actions that must be accomplished step by step by the user in order to succeed in obtaining a result spectrum that characterizes the analyzed sample with a specific surface analysis technique. The underlying simulation engine allows actions that are logically unconnected to be performed in any order. Unlike nuclear power-plant simulators, the protocol simulator does not provide any means to introduce errors or to start from an undefined situation. For now, the simulator prevents the user from doing incorrect actions that may damage the setup, but an industrial version could easily integrate them.

The protocol simulator is composed of three windows, as shown in figure 4. The « Main » window contains the graphical representation of the setup. Each component is clickable, i.e., it can be read or set to a specific value. For example, the state of the accelerator can be observed by just clicking on it. The « Synopsis » window shows the protocol to be followed. Each step can be processed manually (the user performs the action), or automatically (the user lets the computer progress until a breakpoint he has set been is encountered). This allows the user wishing to test a specific sequence of operations to proceed directly to the interesting point. The « Control » window displays the control panel of the currently selected component in the « Main » window.

3. Design methodology

The development of such simulators is intrinsically tedious, due to the tremendous number of possible actions from the user and their consequences on the setup. To handle this problem, we have developped a complementary programming layer on the top of the authoring tool that manages user actions and simulator behaviour. The resulting program is a stand-alone protocol simulator engine that can be applied to any situation involving discrete events. With a simple hierarchical description of all the steps that must be performed in order to reach a specific goal, it is pretty easy to implement a multimedia protocol simulator that corresponds to a given methodology or simulation issue.

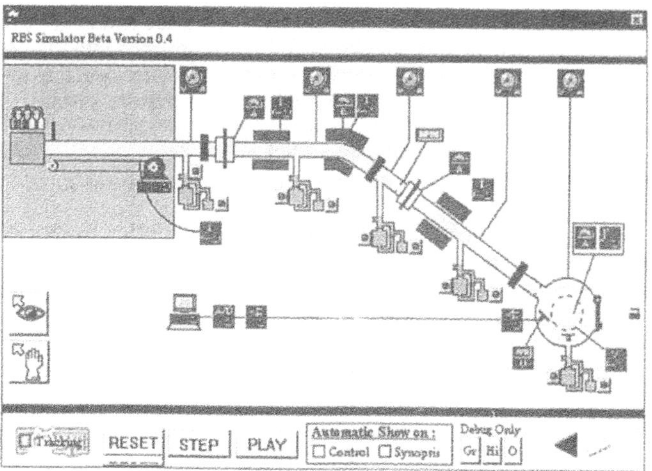

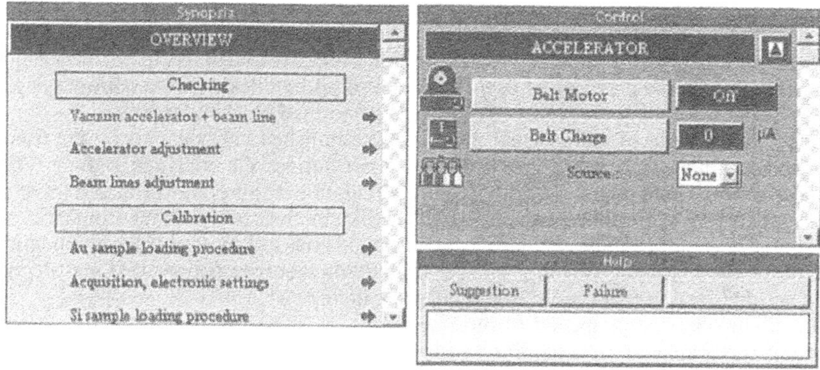

c) The « Synopsis » window d) The « Control » window

Figure 4 : The protocol simulator windows.

4. Results and conclusion

The courses have been evaluated by the MODEM consortium in May 1997. As expected, the MODEM questionnaire on usability of the multimedia courses pointed out that most of our postgraduate students consider the concept of a protocol simulator as a very pedagogical mean to introduce the effective use of complex setups used for surface analysis. The questionnaire also showed that courses can be used at two levels : student and instructor. At the student level, the courses can be considered as stand-alone applications for self-paced learning. At the instructor level, some dynamic parts of the courses such as 3D animations and interactive graphics can be used as efficient pedagogical mean during a traditional postgraduate lecture. We are currently developping the new version of the simulation engine that will allow user-friendly modelling and simulation of any scientific process, from physics to computer engineering.

5. Références

[1] J.A. Leavitt, L.C. McIntyre Jr, M.R. Weller, « Handbook of Modern Ion Beam Analysis » ; Material Research Society, Tesmer & Natasi Editors.

[3] RBS Theory Tutorial, Charles Evans & Associates, http ://www.cea.com/cai/rbstheo/cairtheo.htm, http ://www.cea.com/cai/rbsinst/cairinst.htm

[4] SIMS Theory Tutorial, Charles Evans & Associates, http ://www.cea.com/cai/simstheo/simstheo.htm, http ://www.cea.com/cai/simsinst/caisinst.htm

Session P4

DESIGN INNOVATIONS

ProTest: A LOW COST RAPID PROTOTYPIG AND TEST SYSTEM FOR PCB, ASIC AND FPGA

M. Jacomet, J. Breitenstein, R. Wälti, L. Winzenried, M. Gysel

Biel School of Engineering, MicroLab–I3S,
Quellgasse 21, CH-2501 Biel-Bienne, Switzerland
E-mail: Marcel.Jacomet@isbiel.ch
http://www.isbiel.ch/I3S/microlab/products/protest/e.html

1. Abstract

The test bench methodology helps the design engineer to structure the simulation of his circuit. As showed in this paper, the test bench methodology can further be developed in order to efficiently reuse simulation stimuli and response for the real device under test. As FPGAs are very often used to prototype an ASIC design, an easy switch between simulation and real hardware test is necessary to establish a rapid prototyping design and test environment. Our ProTest system closes the gap between the simulation and the test environment with an easy to use computer–aided–test test tool and low cost test machine.

2. Introduction

The technological evolution in microelectronics has lead to a permanent raise in complexity of VLSI systems integrated on high density chips. Since development and test time should not grow linearly with the complexity of VLSI chips new design and test methods are needed.

At the microelectronics laboratory of the Interdisciplinary Institute of Integrated Systems (MicroLab–I3S) a modern test bench design methodology is used to efficiently design VLSI circuits for research and industry projects. The following steps can be identified:

step 1: Digital systems are designed using a technology independent hardware description languages like VHDL or Verilog–HDL. The digital system is then simulated using the test bench methodology.

step 2: The HDL description is synthesized into a logical format for reconfigurable FPGA chips. The FPGA chips are verified in real time on the low cost rapid prototyping test system ProTest using the already defined test bench.

T.J. Mouthaan and C. Salm (eds.), Microelectronics Education, 161-164.

step 3: The HDL description is synthesized into the target ASIC technology. The ASIC prototype chips are tested on the ProTest environment with the same test bench procedure as used for simulation and verification of the FPGAs.

Synthesizing an FPGA in the ASIC design flow, described in intermediate step 2, has several advantages. This method allows to check for success of the design-for-testability at a much earlier stage. Also specification errors are discovered before producing the ASIC. This does lead to a much improved success rate for a correct design on the first attempt. Thus the presented design methodology is a consequent implementation of the state-of-the art design-for-test approach. The key element of this design methodology is the CAT-ProTest tool and the ProTest test machine with different test adapters for devices under test (DUT) like our FPGAs or ASICs.

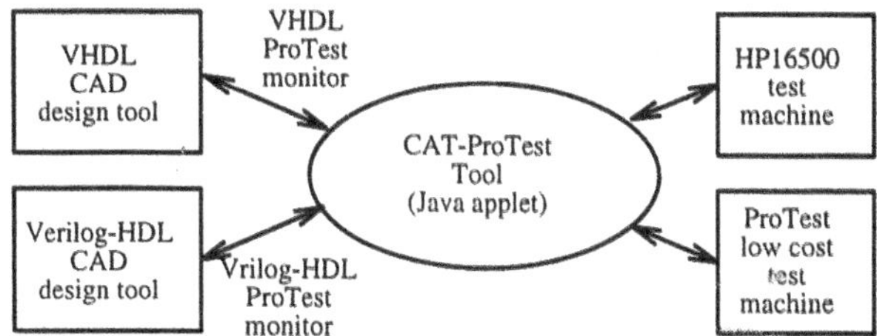

Figure 1. CAT-ProTest tool interfaces CAD design tools and test machines.

3. ProTest System Description

3.1. CAT-ProTest Tool

The Computer-Aided-Test software CAT-ProTest acts as an interface between the CAD simulation environments and the test machines. The CAT-ProTest tool may be used as a standalone tool or may be embedded in a CAD simulation environment. In the embedded user mode, the simulation results from the CAD environments are converted to test patterns and downloaded to the test machine. The DUT test results are taken back into the CAD simulation environment and compared against the simulation in the CAD simualtors window.
The CAT-ProTest tool can also be used as stand alone test development tool, without using any CAD environments. In this configuration the test program can be developed hierarchically with the CAT-ProTest tool by defining clock signals, test patterns and expected responses, test sequences and finally the test programm itself. In stand-alone configuration the test results can be visualized graphically with the CAT-ProTest tool (see Fig. 1).

3.2. Low-Cost ProTest Test Machine

The low-cost ProTest test machine (Fig. 2) is able to generate test vectors and capture DUT responses with a resolution of 100ns. Up to six different clock signals can be generated. The ProTest test machine has 240 configurable signal input, output, bidirectional and clock pins. A library of universal test adapters allows the test of ASICs and FPGAs with standard device sockets.

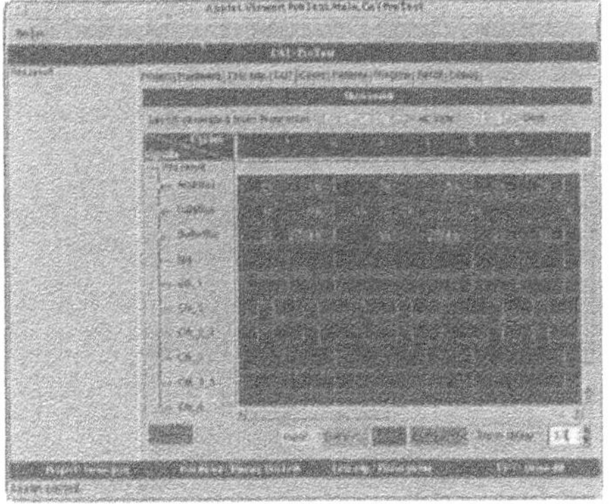

Figure 2. CAT-ProTest tool (Java applet) and ProTest test machine.

3.3. ProTest Monitor

The ProTest monitor is designed as a VHDL and a Verilog-HDL library package. A VHDL test bench consists of a simulation model call, a ProTest monitor call and the test patterns. Every input, output, bidirectional or clock signal of the DUT are defined with a ProTest monitor procedure call. A simple HDL variable is used to switch between the simulation of the circuit model and the comparison between this simulation model with the test response from the real DUT.

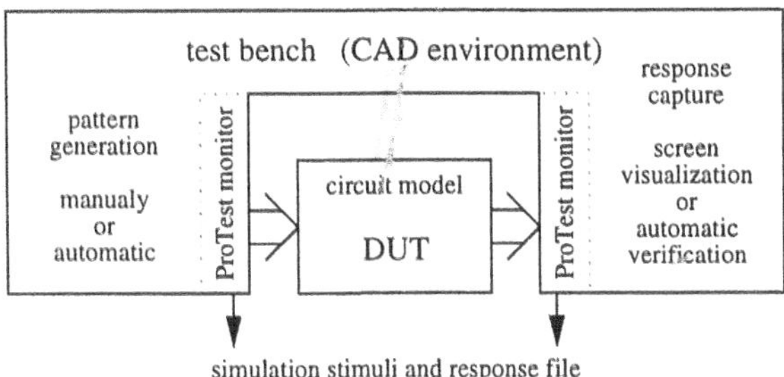

Figure 3. Test bench used for HDL simulation, chip test and result comparison.

4. Test Bench Methodology

Fig 3 shows a classical test bench used to simulate and verify a circuit model described with a hardware description language or with a schematic. The predefined ProTest monitor captures test patterns and simulation responses in a file during regular simulation. The ProTest monitor is activated by a VHDL or Verilog–HDL library call.

Once the digital system is designed and verified, synthesis into an FPGA or an ASIC target technology can be performed. In order to guarantee a rapid prototyping, reuse of the simulation test bench is mandatory. The ProTest environment uses the same test bench as already developed in the HDL simulation and verification phase. Test patterns and simulation results captured by the ProTest monitor during simulation are used to stimulate the real device–under–test.

Design engineers are used to work with CAD environments. Thus design and test procedures are faster if no additional tools have to be introduced to test the DUT. We reuse the test bench developed for circuit simulation for the real tests and result comparisons. Therefore the circuit model is simulated once again and the simulation results are immediately compared with the DUTs response in the CAD environment. In order to perform the comparison of the simulation data with the real chip test data, a simple HDL switch of the ProTest monitor has to be changed from *simulation* to *comparison*. The switch causes the ProTest monitor to load the test result file into the users CAD environment in order to perform the comparison between simulation and DUT response obtained with a test machine.

5. Results and Conclusions

Surveys of HDL users have indicated that the generation of HDL test benches typically consumes 35% of the entire front–end ASIC design cycle. It is clear that the reuse of the test bench for the test of ASICs and FPGAs would significantly reduce the costs of HDL based designs. The presented ProTest system achieves the primary goal to merge design, rapid prototyping and test of ASICs and FPGAs. Student class exercises on design–for–test can easily be executed with the Java based CAT-ProTest tool. Using an application specific test adapter PCB, one or even several integrated circuits can be grouped to form a DUT. In such a configuration the target integrated circuit can be tested in its application environment similar to a bread–board design. The CAT–ProTest software can either be used as a stand alone test development tool or as an easy to handle interface between CAD design tools and test machines.

Due to the ProTest monitor principle described above, all 3rd party CAD simulation tools based on VHDL or Verilog–HDL description language are supported by the CAT–ProTest tool. No CAD simulation tool dependent drivers are necessary. Thus the ProTest system is CAD tool vendor independent. As the CAT–ProTest tool is written in the Java language, the ProTest system represents a highly platform independent rapid prototyping and test environment.

FAST PROTOTYPING for BiCMOS ANALOG INTEGRATED CIRCUITS DESIGN

J. TOMAS, E. RAGBI, P. FOUILLAT and J.P. DOM
Centre Commun d'Enseignement Supérieur de Microélectronique Avancée Aquitaine (CCESMAA)
IXL (CNRS UMR 58 18) - ENSERB - Université Bordeaux 1
351, Cours de la Libération 33404 TALENCE Cedex France
Tél : 33 556 84 65 50 Fax : 33 556 64 28 00
Email : tomas@ixl.u-bordeaux.fr

1 Abstract

At University Bordeaux 1, analog integrated circuit electronics teaching adresses students from 'IUP Génie Electrique et Informatique Industrielle' second and third year, from Licence and Maîtrise Electronique, Electrotechnique et Automatique (EEA) and from DESS Microelectronique (from Bachelors to Masters of Science). Our pedagogic objective is to be able to provide students a concrete example of circuit integration [1], starting from theoretical electronics knowledge to the test of their own designed circuit. In this paper, we detail the actual project proposed in the Microelectronics DESS training courses, focusing on the laser beam writing system developed in our laboratory and on a personal BiCMOS prediffused array designed in AMS BiCMOS 0.8 μm technology

2 Project-oriented course

Dedicated to microelectronics education, we propose a training course which merges the main aspects of analog IC designer's tasks, i.e. circuit analysis, design improvement, layout, manufacturing, die bonding and testing. These tasks are reviewed through various examples as each group of two students has to design specific sub-function.

The theoretical background brought into play concerns the following topics : integrated circuit design, optimisation and simulation of analog cells, layout drawing, thin film technology and photolithography.

The required and used materials are : a workstation, HSPICE electrical simulator, Design Framework II (Cadence), components library of the semi-custom transistors array, wafers, laser writing system, chemical products and standard test equipment.

T.J. Mouthaan and C. Salm (eds.), Microelectronics Education, 165-168.

The planning of this practical training lasts two weeks and requires four teachers.
Let us detail the system configuration for laser beam writing and our developed BiCMOS prediffused array.

3. Laser beam writing system description

The used process consists in depositing the photoresist layer on the wafer, already covered by one metal layer. After exposure and development of the resist, the metal layer is chemically etched and only the desired interconnections remain. The positive photoresist is transformed into a negative one by adding 1.5% Immidazol. A U.V flood exposure is necessary after the laser exposure baking to perform this inversion.
The main steps in this process are listed below :

- Inverted antireflective photoresist (1400-D1-31 Shipley) spin coated
- First bake on hot plate 105°C, during 4'.
- Focused laser exposure using scanning .
- Second bake 130°C, during 6'.
- U.V flood exposure, during 1'.
- Development during 1'10'' in pure developer (microposit developer Shipley).
- Wet etching of Aluminium in a Alu-etch bath.

The laser direct writing system has three main parts : a X-Y translation stage, a laser beam intensity control circuit and the focusing optics. In order to write the desired interconnections, the focused laser beam must be switched on/off and intensity-modulated, synchronised with 2-D movement of the X-Y stage by an acousto-optic modulator [2] as shown in Figure 1.

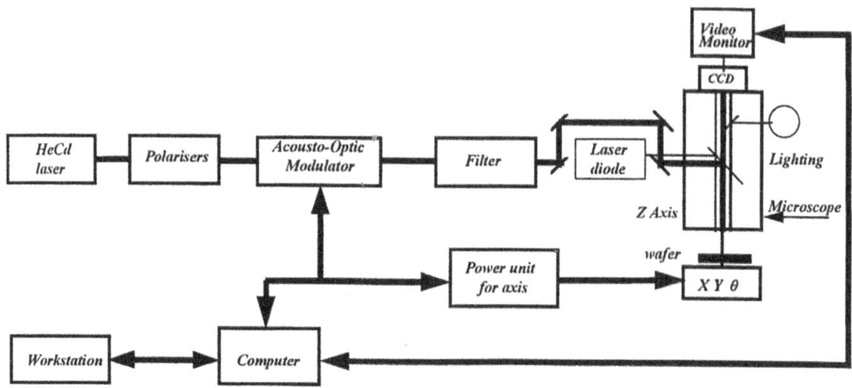

Figure 1 : System configuration for laser writing

Various interconnections of components have been realised using our laser direct writing system. Figure 2 shows the scanning electron microscope (SEM) photograph of NPN transistor.

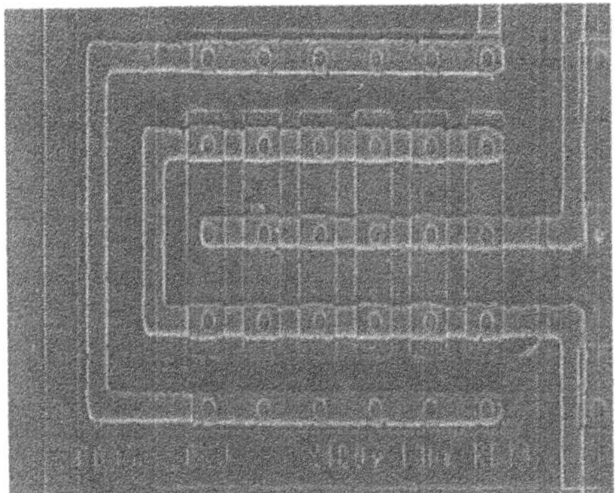

Figure 2 : SEM photograph showing the NPN transistor interconnections

4. BiCMOS Semicustom Array

For our specific application, we have designed a BiCMOS prediffused array with AMS BiCMOS 0.8 μm process. Its more main originality consists in mixing in the same basic cell both Bipolar and CMOS transistor. The array is depicted in Figure 3.

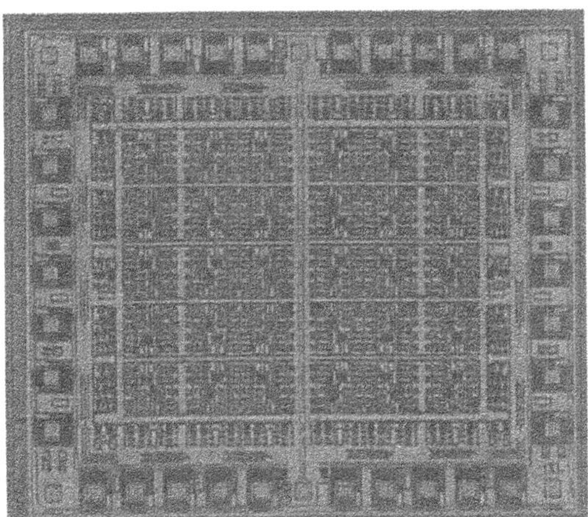

Figure 3 : Semicustom array optical photograph

The chip size is (3.3 mm x 2.8 mm) and is composed of 30 basic cells and 48 I/O pads. Each basic cell contains : 8 small bipolar NPN transistors, 2 lateral PNP transistors, 4 NMOS transistors, 4 PMOS transistors, Poly1/Poly2 capacitor and a large range of values of Polysilicium or Rnwell resistors. A detail of basic cell chip view is depicted in Figure.4. In the external ring, we have also implemented largest NMOS, PMOS, NPN and PNP transistors and capacitors.

The routing layer is the metal 2 layer ; the metal 1 layer is used for supply routing and cross-under.

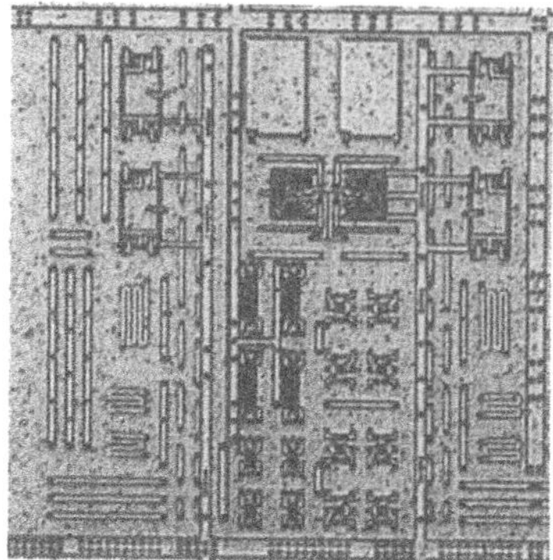

Figure 4 : Optical photograph showing details of basic cell

The laser direct technique is used to customise this bipolar prediffused array in order to validate analog functions. For example, some students have designed the operational amplifier given on figure 5. We have also integrated a 100 MHz PLL.

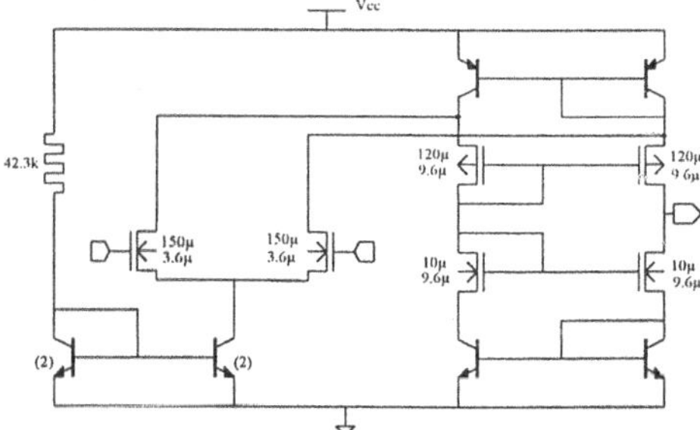

Figure 5 : Schematic of folded cascade operational amplifier

5. Conclusion

This training course has been designed to provide a concrete application example, involving both theoretical and practical knowledge. It introduces the different steps from the design to the measurement. Moreover, it is easy to modulate the training level depending on the students microelectronics background, even if they do not complete all the steps. Thanks to laser writing system, manufacturing delay and cost are hugely reduced. In addition, as the system is independent from the custom array it is well suited to the project future evolutions.

6. References

[1] J. Tomas, J-B. Begueret, Y. Deval, J-P. Dom and J-L. Aucouturier, 'Design and characterisation of analogue integrated circuits - Application to an OPA' proceedings of The European Workshop on Microelectronics Education, Grenoble, France, Feb 5-6, 1996.

[2] E. Ragbi, J. Tomas, G. N'Kaoua, Y. Danto and J-P. Dom : 'Laser direct writing system for fast prototyping integrated circuits', proceedings of ICM'96, Cairo, Egypt, Dec. 16-18, 1996.

[3] J. Tomas, Y. Deval, P. Fouillat, E. Ragbi J.P. Dom and J. L. Aucouturier : 'Design, Integration and Characterization of Analog Integrated Circuits : a Complete Design Flow Dedicated to Microelectronics Education', proceedings of MSE'97, Arlington, USA, JULY 21-23, 1997

THE MIXED-SIGNAL ASIC DESIGN COURSE AT TWENTE

R.J.W.T. TANGELDER, S.H. GEREZ, H.G. KERKHOFF, E.A.M. KLUMPERINK, J. SMIT, H. SNIJDERS, H. SPEEK, H. DE VRIES
School of Electrical Engineering, University of Twente
P.O. Box 217, 7500 AE Enschede, The Netherlands

1 Abstract

In this paper we give a detailed overview of the ASIC design course as it is being given at the Department of Electrical Engineering of the University of Twente. This course covers the complete trajectory from system design via circuit design and actual implementation to testing. Design and testing are not limited to the digital field only, but contain also a substantial analogue and mixed-signal part.

2 Introduction

The current ASIC design course has been introduced in the Electrical Engineering curriculum in the year 1993-1994. The course is compulsory for all EE students and takes about 120 hours. It is divided in four modules spread over the whole year (see Table 1). In our course the students should learn design-oriented thinking. Also there is a strong emphasis on working in groups, with problems like dividing tasks in subgroups, making clear agreements, and working under time pressure.

Module 1	Digital and analogue system design	48 hours
Module 2	Digital circuit design	36 hours
Module 3	Analogue circuit design	20 hours
Module 4	Testing	20 hours
	total	124 hours

Table 1: The four modules of our course

The aim is to give the students a comprehensive and up-to-date course in most of the important aspects of the Application Specific IC (ASIC) design flow. For the development of this course we have used the experiences gained from earlier circuit and IC design courses at our university and the Delft University of Technology. Digital ASIC courses have been given at our University since 1983. However, we have changed and extended the previous courses drastically. While digital system design started previously at schematic entry level, we start now from behavioural VHDL. Also we have included analogue and mixed-signal design into the course. While testing previously only took place at the functional level, we integrated design for testability (DfT), structural digital testing and D/A converter testing theory and practice into the course.

T.J. Mouthaan and C. Salm (eds.), Microelectronics Education, 169-172.

3 System Design

The system, which has to be realised, is a "dialmemo" IC (see *Figure 1*). This IC is a mixed-signal IC connected to a simple keyboard and speaker. Via the keyboard a phone number can be entered and stored, and the IC should produce an analogue dual tone output sequence corresponding to the phone number entered, according to the official CCITT telephone standard.

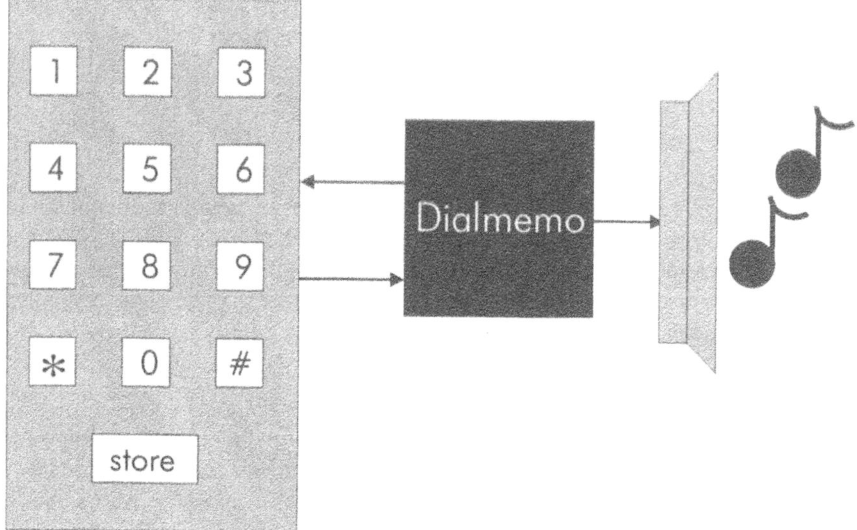

Figure 1: Overview of the dialmemo system

In the first module, the system design takes place. A behavioural top-level VHDL description based on the system specification is made available to the students. It contains a description of the keyboard, a description of the IC itself and a test bench. The IC can be subdivided into a keyboard-scan part, a memory-and-control part and a tone-generator part.
The goal for the digital part of the IC is to refine the behavioural description level to entirely synthesisable VHDL. The provided test bench is used for both pre-synthesis and post-synthesis verification. The system design should be made fully testable using scan paths, to be used in the fourth module. A group consisting of some 8 students each is responsible as a team for the complete design and has to distribute the separate parts among subgroups. For the VHDL simulation we use the Modeltech tool V-System, and for the synthesis towards a testable digital circuit, we use Compass (ASIC synthesiser, Logic Assistant, Test Assistant).
Since the IC should produce an analogue dual tone output sequence according to the official CCITT telephone standard, the tone generator, which is actually a mixed-signal block, has to be examined closely. In this block there are two D/A converters, which are given as library-cells. The D/A converters are used to transform two digitised sine waves to the analogue signals mentioned above. The relation between the numbers of bits used, and the total harmonic distortion (THD) is studied, and compared to the CCITT standard. For the mixed-signal simulation ELDO (embedded in Compass) is used.

ELDO supports analogue behavioural models. Post processing of simulation results is performed with XELGA.
The best among the different student designs is chosen and sent to the ES2 foundry via EuroPractice to be processed in a 0.7 µ CMOS technology (see *Figure 2*).

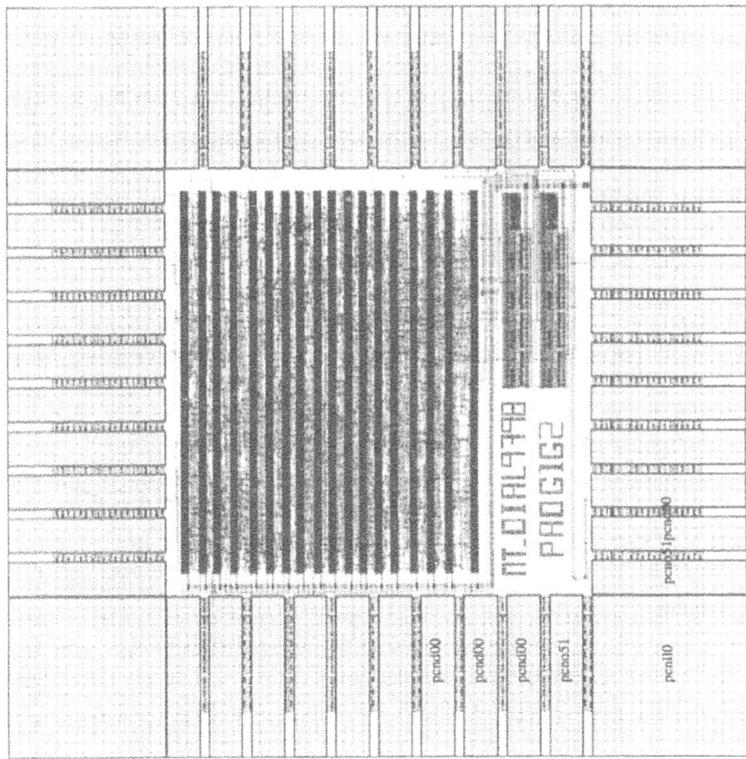

Figure 2: A student realisation of the dialmemo chip

4 Digital and Analogue Circuit Design

While the ES2 foundry is busy with processing, the second and third module takes place. In the first module a library of standard library cells has been used. In both modules the students learn to design these cells. The basic theory of CMOS circuits is explained, as well as an introduction to layout-level design, like the description of the dual-metal layer process and the design rules we use. Translation of a transistor network to layout, layout extraction and circuit simulation without and with parasitic network elements are central parts of this module. In the second module the emphasis is put on digital library cells. The students have to make transistor level schematic of a full-adder and a scan flip-flop, and a 2 input nand and nor cell at transistor and layout level. In the third module analogue cells are studied. The D/A converter used in module 1 is designed in terms of building blocks like current mirrors, current adders and switches. These in turn have to be realised by the students at transistor level. For both modules Compass (Analog Assistant and Layout2) and ELDO are used.

5 Digital and Analogue Testing

In the fourth module the actual realisation of the system is tested. A basic introduction to test pattern generation software and test equipment is given.
The test equipment used consists of an IMS Logic Master ST System, together with a digital multimeter and a digitising oscilloscope.
As an example, first a simple sequential circuit with 6 different states has to be tested. This circuit can be programmed via a digital input to behave as a circuit with a variety of stuck-at faults, which have to be found by the students via structural testing. Once this task has been done successfully, the digital parts of the chip realisation have to be tested also. Since already Design for Testability considerations had to be dealt with at the design phase in module 1, the realisation should be fully digitally testable. For the test pattern generation a local automatic test pattern generation (ATPG) tool is used (which is freely available from our university) together with IMS screen link, to apply the test patterns and to capture the results.
The D/A converter is tested in terms of signal to noise ratio and total harmonic distortion. While putting the parts together in module 1, hardware has been added to induce a fixed digital number, in order to test all the analogue sine waves generated by the tone generator separately. Here the output of the analogue part is analysed with HP-VEE using Fourier transformations, and compared to the simulations done in module 1. Two tests should be done. First the analogue part is tested, while the digital part is turned off. Second the analogue part is tested, with the digital part enabled. By comparing both results an idea of substrate coupling effects can be gained.

6 Conclusions

We have shown how a comprehensive ASIC design course has been implemented successfully at our university. We have the opinion that in such a course state-of-the-art digital and mixed-signal design, synthesis and testing should be included. Therefore, we have tried to be up-to-date in our education and introduced new techniques like VHDL-RTL synthesis and mixed-signal testing.

7 Acknowledgements

We would like to thank all the students, who have been trying out our ASIC design course experiment. Also we have to give acknowledgements to the Technical University of Delft for the distribution of their version of the ASIC design course.

ANALOG INTEGRATED CIRCUIT DEMONSTRATOR FOR ELECTRONIC ENGINEERING EDUCATION

An educational demonstrator for the analysis of analog CMOS circuits

E.FARRES and F.SERRA-GRAELLS
Institut de Microelectrònica de Barcelona. CNM
Campus UAB. 08193-Bellaterra. Spain.

A.URANGA and N.BARNIOL
Departament d'Enginyeria Electrònica.
Universitat Autònoma de Barcelona.
08193-Bellaterra. Spain.

1 Abstract

A demonstrator through the VLSI CMOS 2.5μm double-poly process (CNM25) of the Centre Nacional de Microelectrònica (CNM) only for teaching purposes has been implemented. This demonstrator allows not only the test using normal electronic lab instrumentation of simple analog building blocks and more complex circuits through external connections but also the simulation of these circuits with the LEVEL=2 MOSFET simulation models files provided using an Spice like simulator (for instance the evaluation kit of Pspice).

2 Introduction

One of the main problems dealing with the education on analog integrated circuit design for electronic engineers is the availability to teach the students the importance of the correlation between the design process of one circuit (which takes as a tool the electric simulation of the circuit) and the real performance of the circuit designed. Normally it is very expensive to implement the circuits designed by the students and consequently the students are seldom allowed to compare their simulation pre- and post-layout with the real characteristics of the circuit. Our teaching experience show us that it is extremely helpful for the students to make this kind of comparison. In fact almost all the text books used to teach analog integrated circuits are plenty of simulated examples but seldomly they provide experimental characterization of the circuits [1-4]. We think that in order to validate the simulation is useful to give the student the possibility to make such a comparison.

T.J. Mouthaan and C. Salm (eds.), Microelectronics Education, 173-176.

One of the possible solutions to that problem is to design a circuit which is really implemented; so the device models and the architecture have to be given to the student. In this case the student is able to make the comparison but is only allowed to design a pre-determined architecture. In the design of analog integrated circuits is very important to test several ways to implement the same function and to choose the best for the application required.

Taking into account these considerations a demonstrator with a VLSI CMOS 2.5μm double-poly (CNM25) technology of the Centre Nacional de Microelectrònica (CNM), only for teaching porpoises has been implemented. This demonstrator allows not only the test using normal electronic lab instrumentation simple analog building but also the simulation of these circuits with the LEVEL=2 MOSFET simulation models files provided using an Spice like simulator (for instance the evaluation kit of Pspice is enough). As it is shown in figure 1 it is also possible the implementation of more complex circuits through external connections between the components.

Another important feature for such a practical laboratory is to give a forum for discussion between several groups using the demonstrator. In order to achieve this goal a web page linked to CNM has been designed. All the information related to the chip is included in it: data sheet on-line, some technical applications and also the SPICE models of the transistors used. In these pages it is possible to read/write news and to get feedback both from the users and the providers. In this way a dynamic instrument for teaching analog microelectronic design is provided through the Internet.

3.- Demonstrator

The implemented demonstrator is composed by two matched MOSFET's in different circuit topologies, i.e.: four-terminal characterization device, current mirrors, common grounded source structures and differential pairs. Hence, a wide variety of analog stages can be generated, just by a proper interconnection of these cells, such as: current mirrors (simple, cascode, low-voltage cascode, regulated cascode), inverters, OTA (simple, folded, folded cascode), translinear circuits. **Figure 1 (a)** is a photograph of the implemented demonstrator in a DIL-48 encapsulated chip and **(b)** is a photograph of the demonstrator prepared for direct characterization with normal lab instrumentation. **Figure 2** shows a floorplan including all the available cells in a single dice and its lay-out. Finally **Figure 3** shows the results obtained with a Pspice simulation and experimental measurement with normal laboratory instruments (oscilloscope, function generator and power supply) of a simple differential amplifier. The characterization of Slew Rate and Voltage Gain versus frequency is performed just to show the possibilities with the demonstrator.

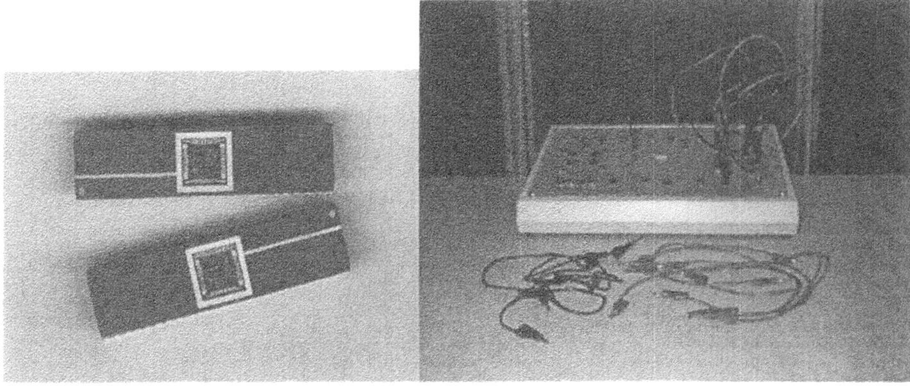

Figure 1 (a) Photograph of the implemented demonstrator in a DIL-48 package and **(b)** Photograph of the demonstrator prepared for direct characterization with normal lab instrumentation

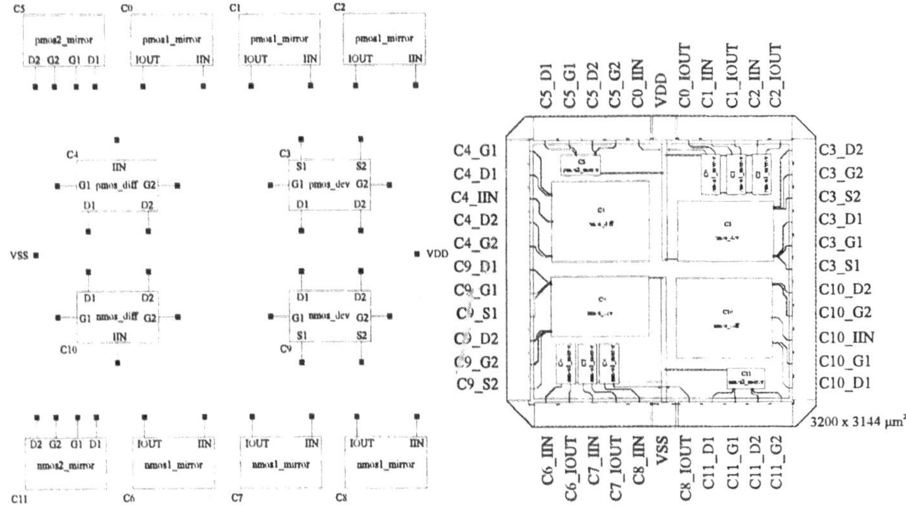

Figure 2: Floorplan including all the available cells in a single dice and its lay-out.

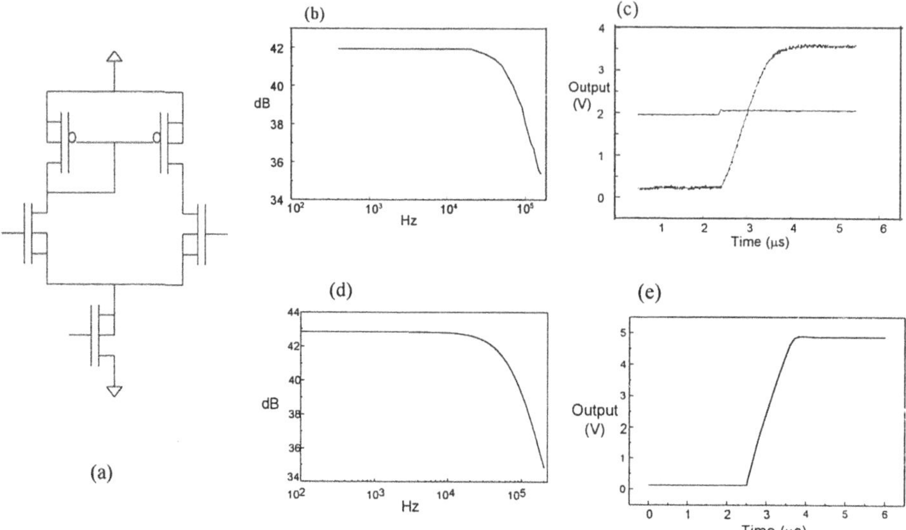

Figure 3: (a) Schematic of the differential amplifier tested; (b)Experimental AC-gain characterisation; (c)Experimental Slew Rate characterisation; (d)Pspice simulation of the ac-gain; (e) Pspice simulation of the SR.

4 References

[1] P.E.Allen, D.R.Holberg. *CMOS analog circuit design*. Ed.Saunders College Publications, (1987).
[2] R.L.Geiger, *VLSI design techniques for analog and digital circuits*. Ed. McGraw-Hill, (1990).
[3] K.R.Laker and W.M.C.Sansen, *Design of analog integrated circuits and systems*. Ed. McGraw-Hill, (1994)
[4] P.R.Gray and R.G.Meyer. *Analysis and design of analog integrated circuits*. Ed. Willey and Sons, 3rd Edition (1993).

i-BUTTON ELECTRONIC IDENTIFICATION TECHNOLOGY: HI-TECH TOOL FOR FINAL YEAR STUDENT PROJECT DEVELOPMENT

M.-T. CHEW, S. DEMIDENKO, B. TOK, D. KOH, J. HON, P.-S. LOH, D. LIM, J. LEE
EC Department, Singapore Polytechnic
500 Dover Road, Singapore, 139651

1 Abstract

Identification Button (*i-Button*) previously known as Touch Memory is a miniature chip-based read/write electronic label packaged into a small button-shaped stainless steel container. It can be attached to virtually any object traveling with it everywhere and providing instant data accessibility. Information in i-Button is read with a momentary contact, and can be updated in the same way while the label is still affixed to the object. The electronic identifier can have virtually unlimited number of possible applications in almost any area of our today's life. In this paper we discuss how the use of i-Button technology can benefit to one of the most important parts of polytechnic student practical training programme - final year project.

2 Introduction

A final year student project is among the most exiting, memorable and favourite student activities during the study at the Singapore Polytechnic. The same is true for numerous polytechnics and high technical schools in the world. The project allows students to apply the knowledge they have learnt throughout the course, to stretch their imagination, to exercise their creativity and to practice the engineering and interpersonal skills acquired within earlier years of the study. At the same time the project is a very important part of the curriculum of electronics, computer and communication engineering specialities taught at the polytechnic.
To satisfy both the roles (i.e., to be interesting and encouraging for students, and to be at the same time significant enough in terms of the practical training programme of the course) the project is required to be on the edge of current technological advances and developments in such areas as Electronics, Information Technology, Computer Engineering, Automation, Communications. It is highly important also to secure that the project is useful and applicable in the social and technological environment of the country, is contributing into improving quality of life, is drawing interest and possible sponsorship from the industry, and, last but not least, is commercially valuable. Such considerations as feasibility to complete the project by a small student team (2-4 students) within a short time frame (6-8 months), and the limitations on the project cost

T.J. Mouthaan and C. Salm (eds.), Microelectronics Education, 177-180.

(usually around US$ 500-1000) should also be taken into account when the technological area and the topic of the project is chosen.
In such a situation to find a good topic for the final year student project is generally not easy task which becomes real 'spring headache' every year for the teaching staff involved (spring is the time for submitting titles for proposed projects).
The main goal of this paper is to briefly introduce an inexpensive hi-tech electronic product that can be applied in almost every area of our today's life (manufacturing, health care, transportation, asset management, postal services, security, maintenance, agriculture, inventory control, etc.), and to discuss its application in polytechnic learning environment. From our experience such harnessing of the state-of-the-art technology can be virtually inexhaustible source of interesting, useful, technologically advanced and commercially valuable student and staff projects.
The name of the product is i-Button (previously known under the name of Touch Memory). It has been rather recently developed in the US, and is at the moment one of the most emerging electronic products in the world.

3 i-Button: hi-tech electronic data carrier

The i-Button is basically a portable memory cell which is packaged in coin-shaped MicroCan to withstand harsh environments. The memory cells can be read or written for virtually unlimited number of cycles. It can replace paper documentation that are difficult to attach to objects and are prone to damage or illegibility. It provides high immunity to electro-magnetic fields, mechanical stress and dirt. No two i-Buttons are the same. Each contains a guaranteed unique serial number that is lasered into the chip at the time of manufacture. This number is a permanent registration code engraved in silicon that provides absolute traceability. The i-Button application software and hardware support tools include an extensive list of computer peripherals, communication and networking means, developed by the chip manufacturer and by number of other companies.

4 i-Button based final year student projects

Application of the i-Button technology to the final year student projects can be extremely wide. Below we present several cases of such applications which have already been developed in the polytechnic.
One of the application areas can be characterized as a personal identification and data carrying tag (badge, card, bracelet, etc.). As every i-Button has a guaranteed unique factory-programmed permanent serial number it can be considered as a globally unique (never duplicated) identifier that can not be counterfeited. Combining this with i-Button's ability to hold some 8000 or more characters of information that can travels with its owner everywhere, be updated instantly, and be password or encryption protected, we get an ideal personal identification document. If however even greater security is required i-Button can be combined with the traditional technologies such as photo, bar code, magnetic stripe, etc. Feasibility of using the i-Button as some sort of

all-in-one ID (keeping personal data, critical medical information, bank and ATM accounts) was studied by student project "Use of "Touch Memory" Chips as basis for Personal Data Card" in 1995.

The other interesting project realised in 1996 academic year was "Smart Guard and Inspection System Using Touch Memory". The main idea of the project is as follows. The i-Buttons (that provide unique serial number, memory and internal permanent clock/calendar within a single component) are attached to some designated locations, e.g., to the objects that should be inspected and premises that should be visited by security guards during their routine patrolling. A small-size hand-held battery-powered i-Button reader/writer is given to each patrolling guard. On the route of patrolling the guard touches the buttons attached to the preassigned locations by the reader/writer. This provides recording at each location the unique serial number of the button, the time and the date of the inspection into the hand-held device. The time and the date of the last inspection as well as some other relevant information (e.g., the results of the inspection) can optionally be recorded into the button itself. Upon returning to the central office (e.g., guard house) the data on the patrolling and inspection are downloaded from the hand-held reader/writer into the main computer. This information then can be stored in the special data base, processed, printed as a formal report serving as an undisputed proof that the guards have visited designated locations and perform their tasks as instructed. One rather important additional feature of this project is that such a guard tour and inspection system could be directly applied in the polytechnic itself improving level of security at the campus.

Access control is one more area for i-Button related projects that are currently under consideration. Such an application can be an ideal solution when featuring high security programmable access control to some restricted areas (buildings, offices, laboratories, etc.) or objects/resources (safes, computers, networks, and so on) is required. The range of complexity of such access control equipment can be very wide - from a rather simple stand alone terminals, providing activation for just a single lock when a valid button is presented, to a very sophisticated networked systems regulating access to a great number of premises (areas, objects, resources) which have different pre-assign levels of security and specified time access zones. The system can also be configured to perform a multitude of additional data collection and other tasks such as: scheduling, attendance, collecting payroll, etc. The data completed with the time, date and i-Button serial number can be recorded and stored for further processing or passed immediately to the system host via modem.

Project "Touch Memory Automatic ID System for Postal Services" was of particular success among our i-Button technology application results. It was featured in our 1997 final year project COMMEX exhibition, and was received favourably among the judges. When Singapore Post required a system check to improve quality of collection/delivery of mails for the public, naturally the i-Button came as a cost-effective and efficient solution. The project was developed in close co-operation with the Singapore Post. It was rated with industrial potential and was displayed in a number of exhibitions and presentations. The main idea of the project is to equip each post box with its own i-Button containing information on it. A postman when doing his round of mail collection will then use the hand-held device (so-called Touch Pen) to record the pick-up time and date plus the relevant information inside the i-Button. Upon returning to the central office, the collected data is down-loaded into PC where it can be stored in a data base, processed and analysed, printed as a formal report. Periodical analysis of the acquired data (scheduling and actual collections/delivery time, average delay/early collections, number missing collections, etc.) allows to take required actions to improve postal collection/delivery services rendered to the public.

5 Project Realization

On the first look i-Button looks rather simple. However, it is in fact a very sophisticated hi-tech product. In order to use it sufficiently a lot of application-important issues specific to the technology, such as memory format, file structure, input/output protocol standards and many more should be learnt and implemented by the staff and students involved in the projects design and fabrication. This could obviously lead to substantial expansion of the project development cycle, that in majority of the final year student project cases is not acceptable. However the projects development can be highly facilitated and accelerated by using not just 'bare' i-Buttons, but so-called Starter Kits. Such a kit provides hardware and software tools for quick evaluation of i-Button using a PC-compatible DOS computer. Since the cost of the kit is not high each project team can easily acquire one-two kits at the very initial stages of the project development or even before the project is started.
The other important issue in order to push forward i-Button based project development is close co-ordination and regular exchange of information, expertise and experience on the matter between project teams. Besides, organising some sort of a practical seminar or workshop on i-Button and its applications for the involved staff and students could be of benefit. Such a seminar would be a good chance to invite highly qualified and experienced overseas specialists working in the area to deliver their talks and practical tutorials on some most troublesome aspects of i-Button technology and its application. At the same time the local participants would be able to present at the seminar their own results sharing by this their experience with the audience.

6 Conclusion

In this paper we briefly introduced an advanced electronic product - i-Button, discussed its possible applications in the final year student projects, and presented examples of such projects. However by writing the paper we target also one more important goal: we consider the paper first of all as an invitation for broader discussion on new technologies that can be implemented in the technical education environment (in the laboratories, student and staff projects, staff consultancy, etc.) making our institutions more competitive and advanced in the fast changing today's world. And we will be glad if this invitation would find response from our colleagues - lecturers, engineers, technologists both from academia and from outside it.

Session D

INTERNATIONAL OUTLOOK

MICROELECTRONIC SYSTEMS EDUCATION IN THE U. S.

D. W. BOULDIN
University of Tennessee
Electrical Engineering
Knoxville, TN 37996-2100
dbouldin@utk.edu

1 Abstract

In the United States, a variety of microelectronic systems courses have been developed during the past two decades. Government and industry actions have played a significant role in shaping these developments inspite of the fact that no coordinated policy existed. These events are chronicled and additional measures are proposed to enhance microelectronic systems education not only in the U.S. but throughout the world.

2 Introduction

Advances in semiconductor manufacturing have fueled an explosive growth in the number of microelectronics-based systems developed during the past twenty years. This growth has created a demand for computer scientists and engineers capable of designing these systems. Design generally involves three tasks:

(1) mapping application requirements into feasible (and optimized) specifications,
(2) implementing these specifications by integrating appropriate microelectronic components, and
(3) creating new components using state-of-the-art technology.

In this process, internal functions must be described as well as the interactions among these components and the external world. Assembly, test and packaging of these components must also be considered. Thus, microelectronic system design generally involves the physical assembly of multiple components with well-defined interfaces.

To respond to this need for microelectronic systems designers, universities have introduced a variety of courses which incorporate design methods appropriate to the components being used. The impact on these developments by government agencies, private industry and universities themselves is described. Additional measures are proposed to enhance microelectronic systems education further.

T.J. Mouthaan and C. Salm (eds.), Microelectronics Education, 183-186.

3 Curricula Development

In the 1980's custom IC courses were introduced at many universities. These were facilitated by four factors: (1) the landmark text [1] by Carver Mead and Lynn Conway which introduced a design method that permitted many of the details of the manufacturing process to be hidden from the systems designer by a well-defined interface to the fabrication process, (2) faculty training courses which were sponsored by the Defense Advanced Research Projects Agency (DARPA), (3) access to silicon fabrication at no charge to universities via the subsidies provided by DARPA and the National Science Foundation (NSF) to MOSIS [2], a broker for multi-project fabrication, and (4) the availability of free software from universities themselves (primarily MAGIC [3] and SPICE from the University of California, Berkeley).

Also, during the 1980's programmable logic devices became available so many universities began updating their introductory logic courses to use these to implement prototypes. The availability of free software from industry, primarily PALASM, accelerated the wide-spread adoption of this technology.

In the 1990's several developments have taken place in parallel. One of these is the widespread use of field-programmable gate arrays. With the help of industrial suppliers who donated or discounted their products (parts and tools), U.S. universities have been able to support laboratory projects in advanced logic design courses. Faculty training courses sponsored by the NSF [4] helped to proliferate these courses. With FPGAs and the associated CAD tools (especially synthesis), students have been able to design more complex systems in less time, thereby enabling them to focus more on design and educational issues rather than the tedious aspects of interconnection.

Two government-sponsored consortia made an impact during the late 1980's and early 1990's. The Microelectronics Center of North Carolina and the Massachusetts Microelectronics Center supported universities in their states and elsewhere with multi-project fabrication and testing, faculty training as well as CAD tool acquisition, installation and support. Although these services proved to be highly beneficial to the associated universities, these operations ceased in light of reduced availability of state funds.

Another influence on the rapid development of university curricular for microelectronic systems was the series of conferences that was sponsored by the NSF [5]. These forums provided for the interchange of ideas and even the sharing of course material. After a lapse of several years, this series of conferences has been revived, but without government assistance [6].

4 Role of Government Agencies

In the United States, the role of federal government agencies (DARPA and NSF) is generally that of a catalyst to initiate or facilitate an area of national interest which might not otherwise be supported. Once the area has achieved a level of independence, government support generally tapers off with the expectation that interested industrial partners will take over. A coordinated policy between government and industry is usually not formulated in an effort to let competitive forces take place in an open marketplace.

A reduction in federal government support of MOSIS fabrication for educational projects is currently planned. It is likely that government funding will be reduced this year to 50% of its present value and then to 25% next year. An effort is already underway to solicit industrial support so that this valuable experience can continue [7]. Almost 200 universities currently use the MOSIS service for educational projects. About 70% of these are for small custom chips which include analog or mixed-signal designs. Almost all of these same schools have FPGA-based projects to support synthesis and digital logic courses. It should be pointed out that NSF did sponsor the development and dissemination of the Analog VLSI Design Resource Kit [8].

5 Current and Future Developments

New courses are being introduced that focus on experimental system building in which rapid prototyping of complete microelectronic systems provides students with the experience of taking complex designs from initial specification to a working implementation. These courses require students to work in teams to specify and optimize systems at a much higher level than in the past. In order that students build such complex systems that involve multiple levels of abstraction, realistic projects must be defined yet prototyping time minimized. To accomplish this, intellectual property (I.P.) blocks are needed to serve as design frames which provide a framework or environment that elevates the functionality of the system and yet helps to manage the project complexity. An international effort to facilitate the sharing of I.P. blocks is already underway [9].

The use of the world-wide web to share information and possibly to provide remote access to complex CAD tools, expensive testers and other services is an exciting possibility. A web site for course listings already exists [10].

Additional improvements in system-level performance and cost will require the designer to focus on timing issues and the interconnection of multiple chips. Multi-project fabrication of multi-chip modules is now being offered to universities via MIDAS [11], a broker sponsored by DARPA.

In the future, there will be an increasing need for computer scientists and engineers who are generalists. Tomorrow's microelectronic system builders must be equipped with a broad education in the fundamentals, a capacity to synthesize diverse information and viewpoints, and an ability to understand and manage extremely complex projects and processes. The trend will be to specify and optimize systems at a much higher level than in the past. This will require high-level software tools, integrated with existing synthesis tools, which will drive the design down through the synthesis hierarchy, to the level of technology and manufacturing processes. The building of higher performing microelectronic systems will require consideration of interconnect technology, power and packaging, device and process simulation, optics and photonics, microassembly, and integration of multiple technologies in a single module or on a board. Teamwork skills and an interdisciplinary approach will be mandatory.

6 References

[1] Mead, C. and L. Conway, *An Introduction to VLSI Systems*, Reading, MA: Addison-Wesley (1980).

[2] Tomovich, C., "MOSIS--A Gateway to Silicon," *IEEE Circuits & Devices Magazine*, vol. 4, nr. 2, pp. 22-23 (March 1988). Also see: http://www.mosis.org

[3] MAGIC Source Code: http://www.research.digital.com/wrl/projects/magic/magic.html

[4] Smith, M., Dillion, J. and M. Zaghloul (eds.), "Field Programmable Gate Arrays in the University," Washington, DC: MIPS Division, National Science Foundation, (1990).

[5] *Proceedings of the 1988-1991 VLSI/Microelectronic System Education Conferences*, Conference Management Services, 407 Chester Street, Menlo Park, CA 94025.

[6] *Proceedings of the 1997 Microelectronic Systems Education Conference*, IEEE Computer Society, http://www.computer.org/conferen/proceed/7996abs.htm (July 1997).

[7] U.S. Committee for Industrial Support of Educational VLSI: brown@engin.umich.edu

[8] Analog VLSI Design Resource Kit: http://www.ece.umassd.edu/ece/studies/vlsi/readme.htm

[9] Intellectual Property Blocks for Design Reuse: http://www.design-reuse.com

[10] Microelectronic Systems News: http://microsys6.engr.utk.edu/ece/msn

[11] MIDAS Multi-Chip Module Brokerage Service: http://www.isi.edu/midas/midas.html

HELPING TO SHAPE MICROELECTRONICS EDUCATION IN CANADA

D. Gale, A. Marsh
Canadian Microelectronics Corporation
Queen's University, Kingston, Ontario, Canada, K7L 3N6
http://www.cmc.ca

1 Introduction

Microelectronics education in Canada is characterized by a diversity of interests and circumstances. This paper describes in general terms some of the interrelated issues facing educators and students against a situation of industry growth that is straining educational capacity for people well-qualified in microelectronic subject areas. Specific details are added for a particular slice of the educational system—the national infrastructure program delivered by the Canadian Microelectronics Corporation.

2 Education and Microelectronics in Canada

A National Perspective: The microelectronics industry in Canada is relatively small but very successful. More than 13,000 people are employed in companies which generated $CDN 4B in revenue in 1995, of which more than 80% came from export markets. These companies represent a significant base of knowledge and experience in microelectronics design and manufacture, and they have strong links to a number of universities and research laboratories. Microelectronics component production is showing significantly greater growth than the rest of the Canadian manufacturing sector.

The Canadian microelectronics industry works through the Strategic Microelectronics Consortium (SMC) to identify strategic and widespread issues in microelectronics. In addition to the need for a state-of-the-art silicon manufacturing facility, SMC has identified a shortage of skilled people as an impediment to the desired growth of the sector. [1] SMC is quantifying how demand for people affects industry growth as a key factor to influence changes in education.

Microelectronic companies are vocal about the use of taxes in education but in general have no direct control and only moral persuasion to influence allocation of resources to their preferred disciplines. There are two major exceptions: i) Facilities and engineering services delivered by Canadian Microelectronics Corporation (CMC), and ii) System, circuit and device research thrusts nurtured by the MICRONET network of centres of excellence. In both these cases Canadian industry provides significant resources and leverages

T.J. Mouthaan and C. Salm (eds.), Microelectronics Education, 187-190.

significant government funding aimed specifically at microelectronics. The primary interest is the development of a strong base of research and education and thereby an increasing number of better qualified students.

3 Influencing Microelectronic Education through Methods and Facilities

A National Infrastructure: CMC is a national not-for-profit organization established in 1984 to provide industrial microelectronic technologies to Canadian universities, both to facilitate world-class research and to ensure a strong source of well-trained graduates [2]. Technologies and services delivered by CMC are similar to those made available through EUROPRACTICE, CMP and MOSIS, and includes some features tailored to the Canadian situation. The portfolio includes, for example:

- Coordinating access to fabrication technologies such as 0.35-micron CMOS (TSMC through PMC-Sierra Inc.), 25 GHz Bipolar (Northern Telecom Limited), and 2.5 GHz Bipolar linear array (Gennum Corporation)
- Negotiating special licensing arrangements or directly distributing selected CAD tools
- Preparing and distributing design kits for the primary CAD tools supported by CMC
- Loaning and maintaining computing equipment and test instrumentation.

Educational programs benefit from this national infrastructure in several ways:

- Intermittently, lecture-based and hands-on training is offered for point tools; these sessions are aimed at students or faculty involved in advanced degree programs.
- Although directed to advanced degree work, many of the licensed tools or technologies (subject to availability) may be used by undergraduate (first degree) students.
- For undergraduate programs the Bipolar array analog technology is available twice per year for courses giving students exposure to the complete design-manufacturing-test cycle (two-week turn-around time from metallization data to 20-pin DIPs for Gennum's technology).
- Documented test examples, with devices, technology files and supporting explanations, are available to introduce techniques and instrumentation.
- Documented design examples, with technology files and supporting explanations, guide first-time users through a digital design flow in a tutorial manner. [3]

Given the demand for highly qualified people and the related interests of Canadian companies to discover more about the educational process, a recent initiative was undertaken by SMC in collaboration with CMC. This intensive "Bridge Camp" was delivered in July 1997 and exposed students to:

- Lectures on analog and digital design delivered in usual classroom settings.
- Laboratory practices in analog and digital design and test; use of tools, design files, and equipment from the national infrastructure.
- A client-specified design project with staged design reviews.
- Almost 20 lectures or case studies delivered by industry representatives covering the business environment, marketing issues, design for manufacturing, packaging, memory design, failure analysis, understanding customer specifications, design for re-use, cost-reduction, design sign-off, and other industry issues.

- Visits to industrial sites.

Influencing Design Methodology, Influencing Microelectronics Education: One way to focus interests, effort and resources is through design flows. A design flow can be viewed as a practical manifestation of design methodology which is at the core of a student's learning experience or represents the know-how underlying product development: a point of intersection of educational and commercial interests.

Many of the items presented in the previous section encompass design flows. They represent moving targets as the national infrastructure evolves in step with industrial interests while delivering technology into the educational environment. In 1997 CMC released a forecast of its technology targets looking forward three years. Within the forecast are specific methodology objectives intended to be compatible with the mix of facilities, manufacturing processes and services. The forecast is itself defined by looking at a global context, for example, the well-known SIA Technology Roadmap [4] and then, with the help of industry and academia representatives, translating this into a national context which is further refined for CMC purposes.

Design flows are defined by CMC working with other practitioners from companies and universities. For example, Figure 1 summarizes the digital flow that, due to rapidly developing deep sub-micron design interests, is now superseding others used in on-line tutorials [3] and the Bridge Camp. CAD tools, technology files for selected manufacturing processes and instructional materials are prepared around the flow. Other flows are under consideration based on the forecast technology targets (which are updated annually). By exploring and understanding the ingredients that comprise a flow one is able to make trade-offs that lead to competitive advantage rather than detract from it. This is becoming part of the learning experience of students using the CMC program and national infrastructure.

4 Conclusion

There is considerable change occurring in the delivery of education in Canada. Although constrained by budgets and working at capacity, the university and college community needs further innovation to make progress on a reasonable scale. This may happen through new forms of collaboration such as industry-academia participation in defining design methods important to achieve technology objectives considered important for microelectronics in Canada.

Acknowledgements

The Canadian Microelectronics Corporation gratefully acknowledges a major funding grant from the Natural Sciences and Engineering Research Council, and contributions of time, effort and technology from several Canadian companies, in particular advice from the Standing Committee on Design Methods and Tools.

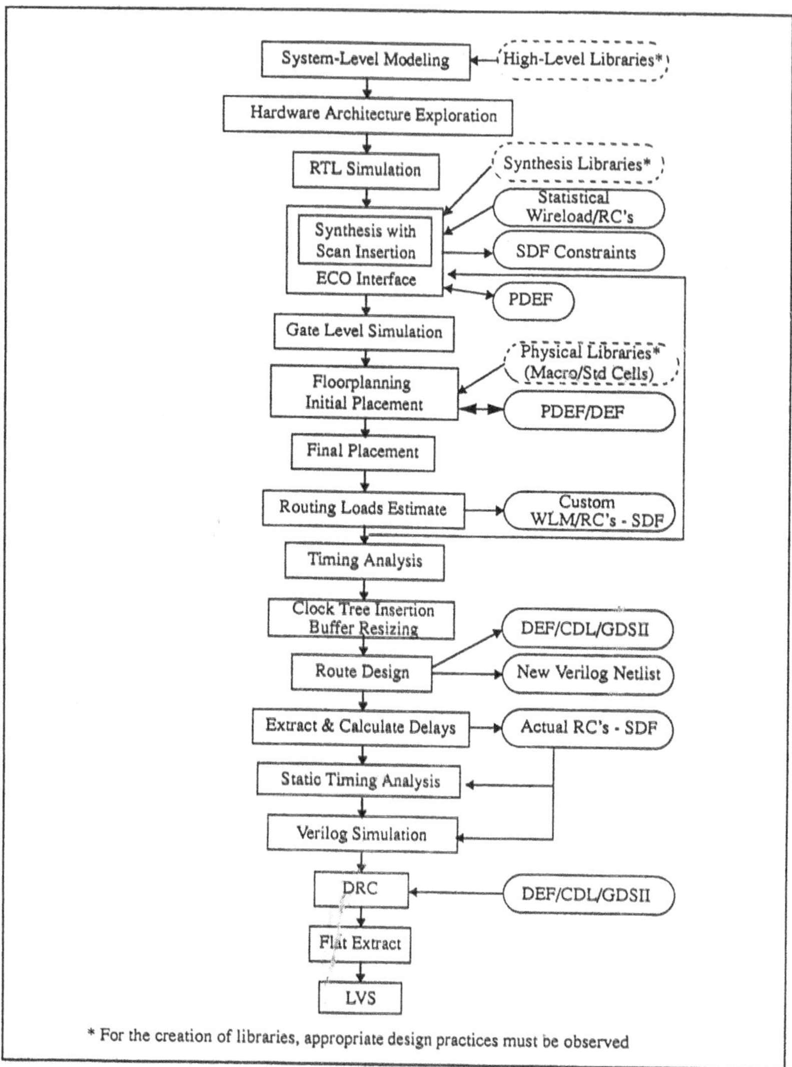

Figure 1 Digital Design Flow (0.35-micron and below)

References

[1] *Canadian Microelectronics Corporation, Progress 1995-97 and Plans 1997-2000*, Canadian Microelectronics Corporation (May 1997)

[2] J. Barby, D. Gale, A. Marsh, I. McWalter, A. Scott, "From IC Design to Embedded System Design: Canadian Experiences and Plans," *Journal of Microelectronic Systems Integration*, Vol. 2, No. 4, pp. 205-221 (1994)

[3] *Basic Digital IC Design Flow Instruction: From RTL Description to Completed Design Using Cadence 9504*, Royal Military College of Canada and Canadian Microelectronics Corporation (November, 1997)

[4] *The National Technology Roadmap for Semiconductors: Technology Needs*, Semiconductor Industry Association (1997)

THE REVIEW OF ENGINEERING EDUCATION AND THE NEW ROLE OF COOPERATIVE EDUCATION

H. B. HARRISON, C-J. PATRICK
Faculty of Engineering , Innovation Centre for Engineering Education
Griffith University, Queensland 4111, Australia.

1 Abstract

This paper provides a brief overview of the recent Review of Engineering Education carried out in Australia. It also presents some of the review findings by way of recommendations, in particular those relevant to cooperative education. These are then put into perspective with regard to an existing cooperative education program namely, the Industrial Affiliates Program (IAP) and a proposed franchise of that program as the ICEE CREAM initiative.

2 Introduction

Engineering Education in Australia has been exposed to a very far reaching review that has resulted in significant recommendations that if implemented fully would change the face of education in that area in Australia. This paper presents briefly the review findings and how they can be used to shape cooperative education programs. We present here a program that builds on the very successful IAP, that has been conducted at Griffith University over the last seven years.

3 The Review of Engineering Education

Late in 1994 the Australian federal governmental Department of Employment Education and Youth Affairs (DEETYA) commissioned a report into Engineering Education on behalf of the Institution of Engineers Australia (IEA), the Australian Council of Engineering Deans (ACED) and the Academy of Technological Sciences and Engineering (ATSE). Two years later, in December 1996, the report in three volumes was handed to the Minister for Education and, through its fourteen detailed recommendations, heralded radical changes for Engineering Education in Australia[1].

The recommendations are summarised in the Table 1 below.

T.J. Mouthaan and C. Salm (eds.), Microelectronics Education, 191-194.

Table 1 Summary of Review Recommendations

1.	Engineers must receive a broader education and be drawn from a wider range of backgrounds
2.	Student intake must be sufficient for Australian Industry to remain internationally competitive.
3.	Engineering courses must have clearly stated goals and equip graduates for life long learning.
4.	Professional accreditation systems must encourage innovation in course content and delivery.
5.	Each University should consider the viability of its engineering school.
6.	Internationally competitive Advanced Engineering Centres must be developed.
7.	Engineering schools must form alliances and facilitate student mobility.
8.	An effective and independent National Centre for Engineering Policy must be established.
9.	Schools and community liaison must be enhanced so that students are more aware of the role of engineers.
10.	The four year full-time course equivalent must remain the minimum requirement, but diversity is to be encouraged.
11.	Staff profiles must balance teaching, research, professional practice and community skills.
12.	Engineering schools must be prepared to collaborate to produce innovative courseware.
13.	There must be greater collaboration between engineering schools and industry.
14.	The sponsoring bodies must take immediate action to implement these recommendations.

Of importance to this presentation are those recommendations that involve industry and in particular that of a consideration of the accreditation system for undergraduate engineering education that has subsequently been reviewed by the ACED in conjunction with the IEA. The IEA being the accrediting body for engineering courses in Australia [2].

In brief, the new requirements for accreditation and the spirit of the review through its recommendations were as follows:

- That students be involved with industry in a more formal manner (previously a loosely defined twelve week industrial exposure was sufficient);
- That staff be in tune with modern industrial practices and seek ways in which they can be more involved with industry;

- That courses be more industrially relevant;
- That industry contribute to modern infrastructure for engineering undergraduates;
- That students be assessed on outputs and achievement of objectives rather than content and inputs.

Currently, the stakeholders (ACED, IEA and ATSE) are actively pursuing the implementation of the recommendations where possible. The following is a response by Griffith University's Faculty of Engineering to address and implement as many of the points touched on above, the concept of which is offered and franchised through the Innovation Centre for Engineering Education (ICEE).

4 The ICEE Cooperative Education Program (CREAM)

Cooperative education is not new to Griffith University. Even before engineering courses were conducted, industrial-based project work was offered [3]. However, with the introduction of a course in Microelectronic Engineering (the first of its kind in Australia) the above approach was formalised and provided as the Industrial Affiliates Program (IAP) [4].

One of the most significant attractions to students of the IAP is that it is incorporated within the timeframe of their normal degree structure and does not extend their study program. Hence the costs of their study program are kept in check. The program also significantly increases students' employment outcomes which of course are of benefit to both the student and the university. The ICEE CREAM take as its base the IAP, which undoubtedly goes a long way to addressing the points raised and summarised in 2 above. Based on the experience and success of this program, the recommendations of the Review will be incorporated into a program with even more industrial relevance, while still maintaining the academic rigour of the present program.

The IAP certainly covers the aspects of formality, detailed reporting and industrial involvement as far as students are concerned [5,6]. This is achieved by proper documentation in planning and conducting the program which is done with an industrial partner over four of the five days in the academic week. The other day is spent on campus reporting to campus supervisors and to their peers as well as attending formal lectures relevant to the program. The program also ensures that by definition it has industrial relevance.

This aspect also results in student access to infrastructure that is not necessarily available on campus. This is a very important consideration for students who are following a Microelectronic Technology path where it is very costly to have even non-state of the art access to equipment [7]. Students can, for example, be placed in a fabrication facility with if not direct access to equipment at least indirect and to personnel who are heavily involved in the industry. Such a placement has exposed students to a real fabrication environment that is not available in academia.

Another area in which the ICEE CREAM concept of the IAP will expand on its present success is by increasing staff involvement. In the ICEE CREAM model the concept is that the staff become involved with industry in a manner somewhat similar to the way students become involved. That is, an industrial partner may present a program that could benefit from the use of an academic staff member and perhaps a student or two and the staff member is encouraged to do this project through financial incentives as well as through staff promotional opportunities. There are many avenues in Australia, at least, to tap into additional funds through training schemes. The federal and state governments have incentive program to encourage training of staff and these are relatively easy to tap into.

Finally, the avenue of assessment of performance is indeed interesting and casts some shadows on traditional means of assessment such as the formal examination process which has been with us since the beginning. The IAP is really a performance based program as it is, but in its present form it has little assessment input from the industrial partner. The ICEE CREAM would seek to enhance this by providing a path way for industry to become involved in assessment. This may be through some panel which has involvement of the industrial partners or other mechanism where the performance results could be somehow standardised and results then allocated.

Performance based assessment is something that is difficult to objectively assess and will be addressed in the ICEE CREAM approach.

5 Conclusion

We have presented briefly the case for review of cooperative education mainly in an Australian university on the basis of an existing, successful course. The concept of a new cooperative education program is available for use at other universities by franchise through the ICEE CREAM program.

6 References

1 *Changing the Culture: Engineering Education into the Future*, The Institution of Engineers, Australia, 1996 (ISBN 858256630).

2 *Manual for The Accreditation of Professional Engineering Courses*, The Institution of Engineers Australia, 11 National Circuit, Barton, ACT 2600 , Australia.

3 D V Thiel , Discussions and Correspondence. 1988.

4 H B Harrison, *A Green Field Challenge*, Proc. of the PRCEE-88 Conference, Sydney, December 1998.

5 C-J Patrick, *A Customer Focus in Cooperative Education*, 10th World Conference on o-operative Education, South Africa, August 1997.

6 H B Harrison and C-J Patrick, *A generally Accessible Co-operative Education Program*, 10th World Conference on Cooperative Education, South Africa, August 1997.

7 C-Patrick and H B Harrison, *Cooperative Education from a User Perspective*, Proc. of the AaeE97 Conference, Ballarat (Aust), December 1997.

MICROELECTRONICS EDUCATION IN JAPAN

VDEC: the center of VLSI design education in Japan

K. ASADA
VLSI Design and Education Center, University of Tokyo
7-3-1 Bunkyo-ku, Hongo, Tokyo 113, Japan

1 Introduction

VLSI Design and Education Center (VDEC) was established in 1996 as an inter-university center at the university of Tokyo for promoting education of VLSI design in Japan[1]. VDEC is a kind of COE of VLSI design education, which covers mainly three functions; (1) providing educational information for VLSI design, (2) providing CAD software tools and (3) supporting implementation of VLSI chips. This paper describes its organization, facilities and activities with future plans.

2 Organization

VDEC has a nation-wide network of VLSI educators. It is actually working as a "network center" composed of sub-centers spreading in Japan as shown in Fig. 1. Though sub-centers are not yet officially established, but volunteers are supporting them. There are totally 12 staffs in VDEC, including two administrative persons and one visiting professor from industry. The other persons are permanent teaching staffs; two professors, three associate professors and three research associates.

As VDEC is an inter-university center, its general policy is decided by a steering committee, members of which are from major national and private universities in Japan. The sub-centers are located at nine major national universities, covering all over the Japan. In each sub-center there is at least one representative professor. At the

Fig. 1 Distribution of sub-centers (9 universities+ VDEC)

[1] Historically, its concept was intensively discussed in early 1980's. Unfortunately it was not realized because of some economical situations in the middle of 1980's in Japan.

T.J. Mouthaan and C. Salm (eds.), Microelectronics Education, 195-198.

end of 1997 fiscal year, there are 124 universities registered as users of VDEC. The total number of registered educators is 331.

3 Facilities

Facilities in VDEC are divided into three categories; network servers and workstations, CAD software and VLSI testing apparatuses. In order to provide CAD licenses and educational information along with verification and storage capabilities for the final VLSI data designed by users, VDEC has network servers with 70 CPUs. Users can access to them at any time via network for the purposes of design verification and the final submission of VLSI designs. Workstations in VDEC are used mainly for training in seminars of CAD tools and VLSI designs.

VDEC has contracts with CAD makers for licenses of CAD tools. The contracts are renewed every fourth year, and now VDEC has licenses of Cadence, Synopsys, Avanti and Mentor. These companies are providing enough numbers of licenses to be used in all the Japanese universities, namely 500 - 1000 licenses for each category of CAD tools.

Testing apparatuses are set at VDEC as well as sub-centers, so that users can easily access to their local sub-centers. VDEC and some sub-centers have also FIBs for fixing VLSI chips when some malfunctions are found in testing.

4 Activities

4.1 BASIC ACTIVITIES

As described above VDEC has three missions. VDEC is providing text books in forms of on-line and papers, organizing mail groups of CAD users, tester users and chip designers, and planing seminars of CAD tools and VLSI designs. Seminars are held not only at VDEC but also at sub-centers for users' convenience. These activities are going on so as to keep the information and the contents fresh.

Users can use CAD tools provided by VDEC on requests basically without charge. Applications of CAD tools are done via VDEC home page. (http://www.vdec.u-tokyo.ac.jp) Users have to get their ID numbers and passwords in advance to make applications of CAD tools. In responding to the applications, CAD software is sent to the applicants usually in CD-ROMs, which are then installed into their workstations. Licenses are provided via network from local sub-centers or directly from VDEC depending on situations. The license is controlled based on IP addresses and workstations' IDs.

For chip implementations, VDEC is now providing three technologies; CMOS 1.2 um by Nippon Motorola, CMOS 0.6um by Rohm and CMOS 0.5um by NTT Electronics. Fabrication runs of each technology are twice a year. Chip sizes along with IO pads are standardized as shown in Table 1.

Table 1 Chip Implementation Technologies

TECHNOLOGY	CHIP SIZE	PRICE (x1000YEN)
NIPPON MOTOROLA CMOS 1.2um Poly2 Metal2	2.3mm x 2.3mm	66.5
	4.8mm x 4.8mm	230
	7.3mm x 7.3mm	451
NTT ELECTRONICS CMOS 0.5um Poly1 Metal3	2.3mm x 2.3mm	150
	4.8mm x 4.8mm	600
ROHM CMOS 0.6um Poly2 Metal3	4.4mm x 4.4mm	350
	8.9mm x 8.9mm	1,100

Technology files including standard libraries required for CAD tools are provided from VDEC in a form of either on-line files or papers. Before getting this information, users are requested to agree the duty to protect privileged information regarding to the specific technologies. Though the CAD tools are basically free of charge, the chip implementation cost is charged to users as listed in Table 1. For each design 10 packaged samples along with 10 bare chips are delivered about three months after deadlines of design data submission.

Fig. 2 and 3 shows history of the chip implementation at VDEC and a pilot project prior to VDEC. Fig. 2 shows increase of designs implemented at VDEC and Fig. 3 shows the total chip area in mm^2. In 1994 and 1995 before VDEC had been established, a pilot project was running for researching a possibility of chip implementations organized by university people. Though both the figures show exponential growth of chip implementations, we are estimating this trend will saturate at about a double or triple level of the current status in 1997, based on the total number of Japanese educators in VLSI design.

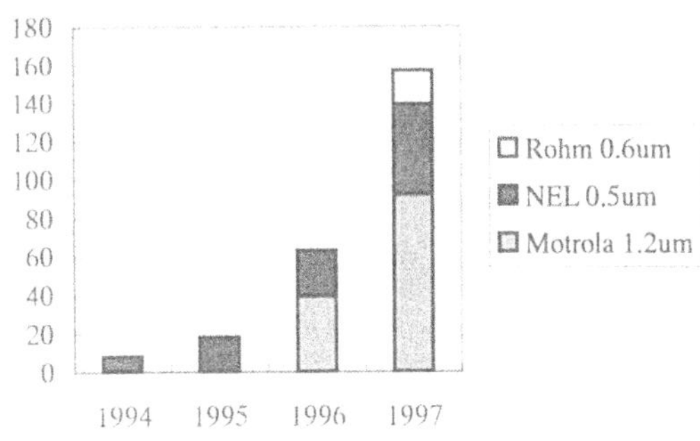

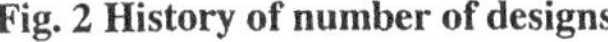

Fig. 2 History of number of designs

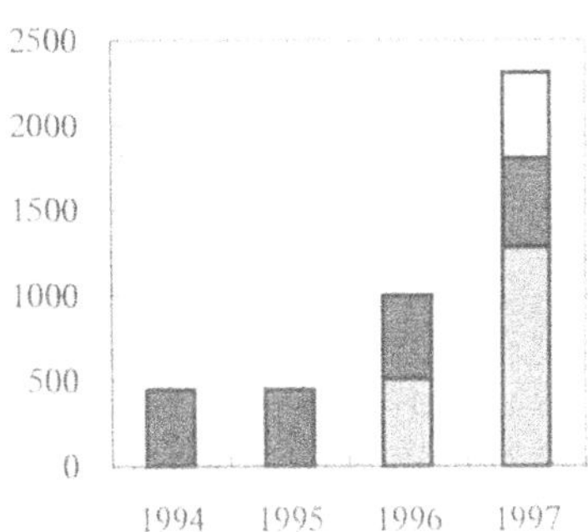

Fig. 3 History of total chip area

4.2 PILOT PROJECTS

VLSI technologies are still growing up in terms of integration density and performance of circuits. We have to continue to renew the implementation technologies, CAD tools and design methodologies to be used for VLSI design education. For this purpose

VDEC is promoting the following projects:

(1) Pilot projects of testing new implementation technologies and CAD tools,
(2) Development of text books for VLSI design and CAD tools, and
(3) Development of model curricula using VDEC functions.

For example Rohm's 0.6um technology was tested in 1997 as a pilot project, where eight major universities participated and several kinds of design methods and libraries were tried and evaluated. Results of this project are the basis of VDEC chip implementation services using the Rohm CMOS technology.

In 1997 we also developed textbooks for CAD tools. Using these books a student can study outline of CAD tool functions by himself within about 2 days for each topics, namely logic synthesis and simulation by HLD, automatic placement and routing along with back-annotation, interactive circuit and mask design.

Tokyo institute of Technology, one of sub-centers, tried to develop a model curriculum for an undergraduate student class, where HDL descriptions were first simulated then synthesized targeting FPGAs. After all the students of the class took part in the FPGA implementations, selected students went up to a full custom implementation course using the VDEC 0.5um CMOS technology. The final designs are assembled as a multi-project chip and fabricated by VDEC. The results and experiences obtained through the above projects will be fed back to all the educators in Japan from VDEC.

5 Future plan

In 1998 we will start a gate-array implementation service using LPGA (Laser Programmable Gate Array) by Sony/ChipExpress for a quick turn around time of one or two weeks. We will continue the pilot projects mentioned above for new targets, further fine and multi-functional technologies and new tools. In the pilot projects we will focus on development and distribution of IPs (Intellectual Properties). For promoting these activities we have a plan to establish a "IP contest" and a "IP Award".

Japanese government will start a system LSI implementation center for venture companies. VDEC will also cooperate with this new center for stimulating each other, though the mission of VDEC is limited to promotion of VLSI design in universities.
We have a desire to "officially" establish sub-centers in several years. The final target of VDEC activities is to establish a network society for VLSI design educators and students as a base of VLSI education in Japan.

Acknowledgement: We appreciate CAD makers, VLSI foundries, and Mask makers for support of the basic activities. We also appreciate Semiconductor Technology Academic Research Center (STARC) for support of the pilot projects.

References:

[1] K. Asada and K. Hoh, Proceed. of the Asia and South Pacific Design Automation Conference 1997 pp.365-369,Jan.28-31,1997,

Session P5

MULTIMEDIA IN MICROELECTRONICS EDUCATION

TEACHING BASIC ELECTRONIC SYSTEMS WITH MULTIMEDIA

Dr. E. Zysman & Prof. M. Declercq

EPFL – LEG CH - 1015 Lausanne Switzerland

Abstract

Teaching the design of basic systems is a fundamental and difficult part of an electronic course. Due to the number of components, the interaction between blocks, perturbations and performances become critical constraints here. This paper describes a methodology and an hypermedia solution developed at the LEG - EPFL (Electronics Laboratory of the Swiss Federal Institute of Technology). This solution assimilated to a virtual lab shows a good complementarity with traditionnal methods. The difficulties of such a project will be presented and a first feed-back will be analyzed.

1. Introduction

Before the introduction to this chapter, students used to analyse or calculate their circuits with different models and laws presented during the lectures. Quantitative previsions were possible. In this part, the number of components and their interactions do not allow a simple calculation. The behavior of the circuit becomes crucial in terms of frequency, stability, linearity, precision, etc....

Simulations can partly reduce the impact of this difficulty; moreover, their repetitive use is beneficial to identify different problems. Nevertheless, simulation doesn't imply understanding. We must then develop an intuitive vision of the circuits. A qualitative approach is then necessary.

2. Methodology

The method is intended to conciliate three constraints which are the global understanding, the intuitive approach and the critical mind. To satisfay these criterias, qualitative analysis and comparisons must complete a traditionnal (quantitative) calculation.

We must then mix various levels of simulations: accurate SPICE-like simulations, simulations obtained with the models proposed during the lectures and more qualitative simulations envolving trends instead of values.

Understanding interaction between blocks suppose a hierarchical vision in terms of functionnal blocks or analog elementary cells. It is thus necessary to propose a structural view where cells can be identified and their local interactions be observed, suggesting and justifying improvements.

T.J. Mouthaan and C. Salm (eds.), Microelectronics Education, 201-204.

3. A multimedia solution

The study of an electronic system starts with a first hierarchical solution representing its functionnal decomposition (fig. 1). The limits of each basical function are evaluated, and different improvements are proposed. This method brings the student to set of circuits including industrial circuits. A global comparison in terms of performances is practical too.

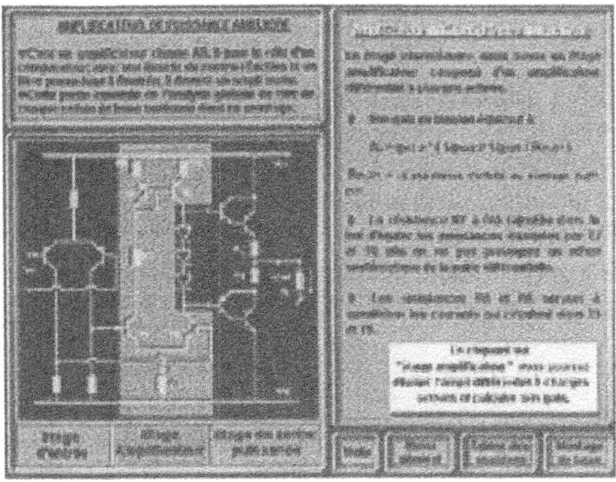

fig. 1: Functionnal decomposition

To help understanding circuit behavior, we propose a qualitative simulation showing the variation trends of each node, as a consequence of a given variation of the input signal (fig 2). An accurate simulation with SPICE is possible and will confirm these trends.

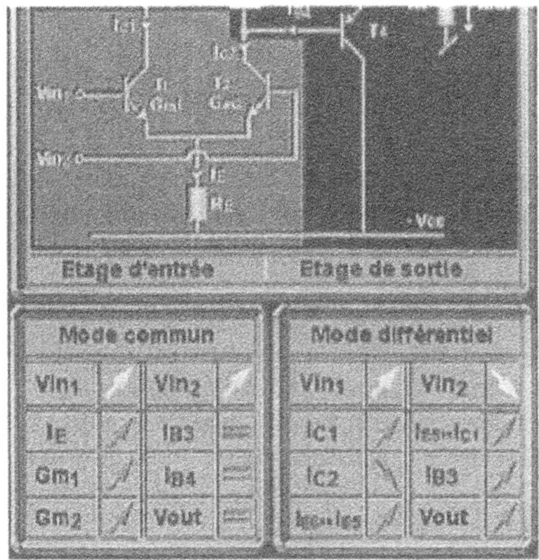

fig. 2: Qualitative simulation

Other qualitative visions will emphasize the relation between an audio amplifier features like the non-linearity effect, the distortion rate, the saturation, the power.

The qualitative approach is an essential feature in the project and not only in terms of simulation. A new chapter dedicated to the amplifier classes present a global vision of the problem. Two lists present respectively the main amplifier classes and various families of applications . When an amplifier is selected, its main features are displayed and some suitable applications are illuminated with colors adapted to the usefulness of the amplifier. The reciprocity is verified when a specific application is selected.

Each electronic system is proposed with a suitable structure allowing to identify and select functionnal blocks or analog elementary cells. Each selection is associated to a context-sensitive aid (fig 3) justifying its use and the problems it generates. From the cells the user can have an accurate vision of the elementary cells (fig. 4). Their functionnality is explained and some experiences validate their role. Here too, we have qualitative simulations based on the models seen during the lectures. Circuits can be simulated with SPICE too. A part of the course material is proposed here to remind some informations.

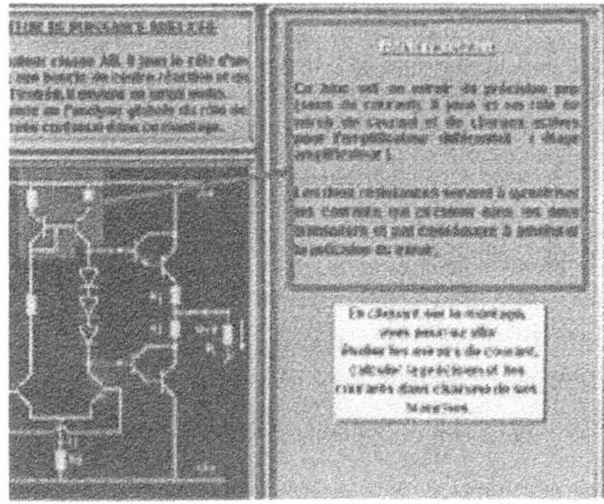

Fig 3: Context-sensitive aid

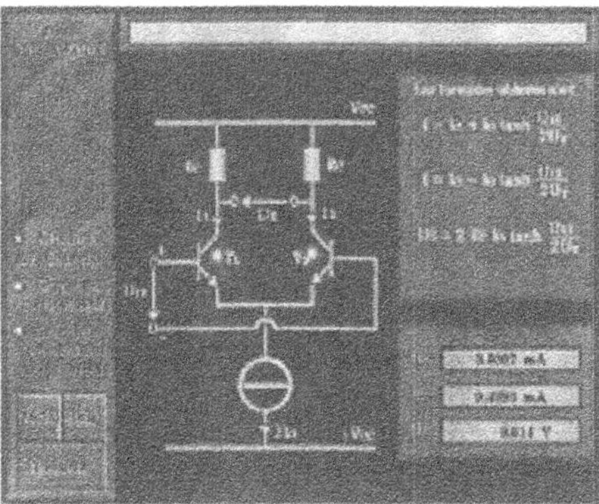

Fig 4: Elementary cell features

4. Advantages of this method

Comparing to traditionnal method, Hypermedia shows many advantages. We assume the principle that clear information and a nice tool presenting it, is a good way to transmit a message.

Like in the laboratories, the circuits can be parametrized. The parameters can be values, but also trends which are not possible in labs. The repetitivity of the experience is important to scan a large range of values.

Interactivity in terms of links is here very interesting too. The permanent and contextual aid is a major advantage, allowing to visualize a complementary information or an index of links.

Animations of physical concepts are also necessary to complete the global view of the students.

Not objective constraints, like esthetic, ergonomy, semiology, are crucial in multimedia. These aspects must be explored to improve the product, facilitate its use and the pleasure to use.

5. Complexity of the project

The development of the Basic electronic systems chapter allows, at this stage, some evaluations. The effort was engaged in various domains:

- 20 % for the scenario, including usefulness evaluation, creativity and feasibility
- 35 % for the Graphism
- 5 % for the ergonomy and screen organization
- 5 % for the integration and the management of the different ressources.
- 40 % for programming animations and interactivity.

1'000 hours were necessary to develop the equivalent of 12 hours of didactical activities.

Obviously, the result does not cover the whole discipline.

6. Conclusion

A first demonstrator based on the power amplifier chapter and the different amplifier classes is realized. In the near future, study of OTA and A.O. circuits will be proposed.

This product corresponds in fact to a second version. After a first feed-back, the analysis showed a desapointment of the students owing principally to the austerity of the application . The ergonomy and the esthetic of the application was completely reviewed. The product is now very well received by the students. The objectivity of the message can now be fastly evaluated, and the first analysis show a great interest.

Owing to the limited part of the course and the effort engaged, we are faced with a fundamental question: Does such an effort worth the trouble? In the future, initiative for educationnal multimedia support will need a bigger structure including various academic groups and industries. Cooperation between countries must be encouraged two. Actually, the language will not remain a restrictive feature for training. As electronic industries turn supranational, training will turn supranational too.

EDEC - A COMPUTER-BASED TEACHING SYSTEM FOR ELECTRONIC DESIGN EDUCATION

P. L. JONES
The Manchester School of Engineering
The University of Manchester
Oxford Road
Manchester M13 9PL
United Kingdom

1 Introduction

The Electronic Design Education Consortium (EDEC) was established in 1992 when eight universities were allocated £1M jointly to develop computer-based learning material suitable for use in the education of electronic engineers and computer engineers. Altogether some 70 modules have been developed representing 150-200 hours of courseware. The material is grouped under four main headings.

1. Electronic Circuit Design
2. Digital Design
3. System and High Level Design
4. Testing and Design for Testability

The courseware complements traditional methods of teaching and learning such as lectures and tutorials and is primarily intended for use by students as a self-study learning aid. It can also be configured for demonstration purposes within lectures, tutorials and laboratory sessions.

The latest release of the EDEC portfolio (version 3) is well-matched to the Windows 95 environment and in addition to general improvements contains some material which was not available in version 2. Distribution is on CD-ROM with minor updates and support via the World-Wide Web. Up to date information about EDEC and a detailed list of module titles can be found on the EDEC Web pages together with ordering and pricing details at

http://edec.brookes.ac.uk

By the end of 1997 there were 31 licensed university sites in the United Kingdom and a further 10 licences at overseas higher educational institutions.

T.J. Mouthaan and C. Salm (eds.), Microelectronics Education, 205-208.

2 Customisation of modules

The EDEC framework incorporates a reconfiguration mechanism which gives teachers the freedom to select which sections and chapters are to be included in a module when it is presented to students. External material (text, graphics, video and sound clips etc) can be linked into the courseware using the framework's 'media link' facility.

Apart from this reconfiguration mechanism the courseware content is not however editable.

3 The EDEC courseware syllabus

The following sections contain brief descriptions of the courseware available in each of the four main subject groups.

3.1 ELECTRONIC CIRCUIT DESIGN

3.1.1 Course and level:
Undergraduate degree courses in Electrical and Electronic Engineering; some of the material is also relevant to degree courses in other branches of Engineering, Computer Science and Physics. Mostly first and second year level, with some more advanced (third/final year) material.

3.1.2 Brief description:
Teaching and learning material covering: Analogue and digital circuit design at the level of circuits and sub-systems using operational amplifiers, transistors, diodes and discrete passive components; relates to low-level CAD tools such as SPICE. Integrated circuit fabrication technology is briefly introduced.

Number. of hours of teaching material: Approximately 40 - 50 hours.

3.2 DIGITAL DESIGN

3.2.1 Course and level:
Undergraduate degree courses in Electrical and Electronic Engineering; some of the material is also relevant to degree courses in other branches of Engineering, Computer Science and Physics. Mostly first and second year level, with some more advanced (third/final year) material.

3.2.2Brief description:
Teaching and learning material covering: Introduction to digital data representation and number systems, combinational and sequential logic design, state machines, microcomputer architecture and ASIC design. Presents an approach to design based on sound theoretical knowledge allied to practical considerations of manufacture.

Number. of hours of teaching material: Approximately 50 hours.

3.3 SYSTEM AND HIGH LEVEL DESIGN

3.3.1 Course and level:
Undergraduate and postgraduate degree courses in Electrical and Electronic Engineering and Computer Science. Mostly advanced level material (third/final year and postgraduate).

3.3.2 Brief description:
Teaching and learning material covering: Specification, architectural design, decomposition and synthesis, centred on the use of VHDL; support material is provided to back-up that contained in the main courseware modules. This includes extended explanations, synthesis restrictions on VHDL, simulation characteristics, glossaries and reference material on the use of VHDL. It is expected that users will already have access to conventional CAD tools for logic synthesis and high-level system design through initiatives such as EUROPRACTICE. Such tools are necessary if students are to gain maximum benefit from the substantial amount of practical work that is centred on VHDL.

No. of hours of teaching material: Approximately 30 hours.

3.4 TESTING AND DESIGN FOR TESTABILITY

3.4.1 Course and Level:
Undergraduate and postgraduate degree courses in Electrical and Electronic Engineering and Computer Science. Mostly advanced level material (third/final year and postgraduate).

3.4.2 Brief Description:
Teaching and learning material covering: An understanding and appreciation of how to apply the principles and established procedures for testing and designing testable electronic systems, and the concept of testing and design for testability being as integral to the design process as considerations regarding performance and functionality. Overview of circuit failure modes, the testing process and types of Automatic Test Equipment (ATE). Methods for Automatic Test Pattern Generation (ATPG), fault modelling and fault simulation. Design for testability techniques. It is expected that users will already have access to appropriate CAD tools through initiatives such as EUROPRACTICE. Although the availability of CAD tools is not essential it considerably enhances the value of the courseware.

No. of hours of teaching material: Approximately 50 hours.

4 Future plans

As a result of a collaboration with the five Polish universities involved with the TEMPUS COHERENCE Joint European Project (S_JEP 07648-94) completed in August 1997, EDEC are now distributing to subscribers to version 3, evaluation copies of the TEMPUS courseware developed on a range of topics including, computer architectures, computer networks, computer graphics, VHDL, reliability and neural networks. The aim is to include modules which receive favourable evaluations as fully integrated and supported modules in the next release of EDEC.

MULTIMEDIA VHDL LEARNING SYSTEMS: AN OVERVIEW

M. KOCH, D. TAVANGARIAN
University of Rostock, Department of Computer Science
Albert-Einstein-Str. 21, 18059 Rostock, Germany

1 Abstract

This paper gives an overview of approaches to teach VHDL design using multimedia learning systems. For this reason we compare commercial and university tools under different aspects. The main aspects are the structure of the whole system (integration of text, audio, animation, lessons and exercises), the use of simulation and synthesis tools within the system, the usability on-line and off-line and the complexity of the exercises/examples.

2 Introduction

What are the goals of a VHDL (Very high speed integrated circuits Hardware Description Language) learning system? VHDL is a complex hardware description language [1]. Therefore, one objective of the VHDL learning systems is to make the student or engineer familiar with the syntax and semantic of VHDL and to provide assistance to problems concerning the language itself. The second objective is to give a guideline for modeling complex designs for simulation and synthesis on different levels of abstraction. This is very important, because first of all VHDL has been developed as a language to model and simulate complex integrated circuits but not for a synthesis of such circuits. Especially, the possibility to use abstract data types to model a system on a higher level of abstraction simplifies the simulation of a complex system, but most of these models are unusable for synthesis. It is necessary to make the student familiar with a modeling style appropriate for the used level of abstraction and optimized with regard to synthesis. And last but not least, the third objective is to make the engineer familiar with special tools of an EDA (Electronic Design Automation) company. This is necessary because the different tools follow different philosophies and because of the complexity of the tools, it is necessary to focus on the tools used daily. Hence the structure of a learning system and the used tools determine the usefulness of a system for the described purposes.
In the following different approaches of VHDL learning systems are explained in detail.

T.J. Mouthaan and C. Salm (eds.), Microelectronics Education, 209-212.

3 Structures of VHDL learning systems

The basic learning systems for VHDL are based on multimedia authoring systems for CBT (Computer Based Training). They are available on CD-ROM (off-line) and/or over the WWW (World Wide Web, on-line). Depending on the environment, the systems include lectures with exercises based on animations and e.g. multiple choice questions or additional exercises are possible using special tools.

The advantage of the CD-ROM systems is the independence from a network. This is especially useful if you want to do some VHDL lectures at home or generally not in specially equipped rooms. These courses are concentrated on the philosophy of VHDL. The main topics are the modeling in VHDL and the underlying structures to model, simulate and synthesize circuits in VHDL. For user defined exercises the systems need additional tools.

The on-line systems are comparable with the CD-ROM systems with the difference that the lectures and exercises are downloaded over the network. This allows simple updates of lectures, software and exercises.

The systems are supported with audio. Especially the Doulos VHDL PaceMaker system [2] uses high quality audio to explain the main topics on each page. First, the explanatory text of a page in the system is grayed out and the topic is explained using graphics, animations and source code examples together with audio. After the oral explanation, the user is able to deepen in the context using the now visible text (see *Figure 1*). In other tools, e.g. in [3], audio is used to explain special topics or as additional comment. Audio is a good completion within a CBT system but has two major drawbacks: First, the use of audio in computer labs, e.g. students labs at our university, is difficult, because the students listening to a course disturb other students doing their exercises. Second, using the system over a WAN (e.g. via modem at home) high quality (large) audio files result in long waiting periods for downloading the data.

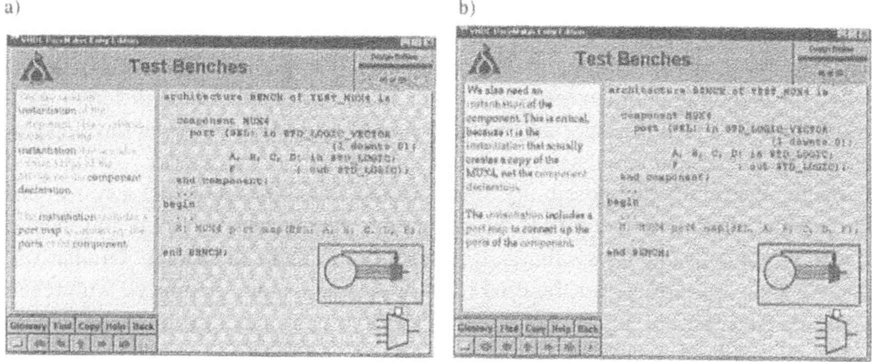

Figure 1: Example: VHDL PaceMaker a) during audio, b) after audio

Figure 2 shows the different methods used for co-operation between the learning systems and the simulation and synthesis tools. The environment necessary to use the system (PC, Workstation, stand-alone (off-line), network (on-line), etc.) is mainly determined by the tools. We distinguish between three major categories: systems without tools, systems with integrated and systems with separate tools.

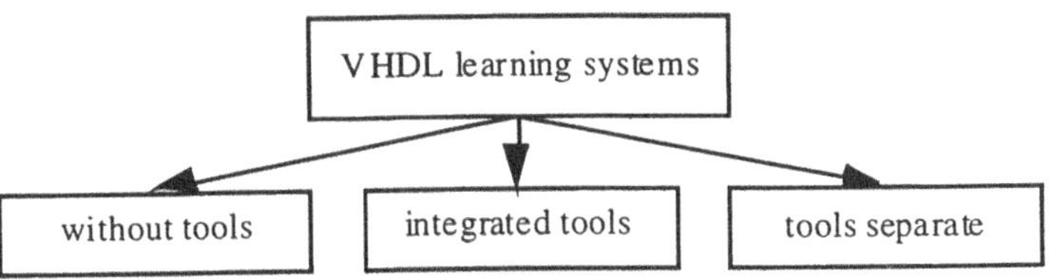

Figure 2: Different structures of VHDL learning systems

3.1 SYSTEMS WITHOUT TOOLS

The possible levels of exercises within the stand-alone CBTs are limited, because the exercises need to be integrated in the system for a stand-alone use. These exercises reach from simple questions to complex problems, e.g. in the Esperan MasterClass [4] and the Doulos VHDL PaceMaker system. The disadvantage of this approach is that the user is not able to use personal examples.

3.2 SYSTEMS WITH INTEGRATED TOOLS

The second category are learning systems with integrated tools. In contrast to the systems in the first category, these systems support the modeling, simulation and synthesis of user defined circuits. So the user is able to gather experience writing his own models, compile, simulate and synthesize them. These systems are mainly browser-based, i.e., the course is transferred over a network using browsers like Netscape or Microsoft Explorer, all calculations are done on a remote compute server. For this, the VHDL source code is edited local, e.g. within the browser, and is sent to the WWW server. The server distributed the tasks over the compute servers running the complex EDA tools. The results are transferred back to the user. The complexity and the user interface of the EDA tools is hidden from the user to concentrate on VHDL. *Figure 3* shows an example of some experiments in this area done at our university.

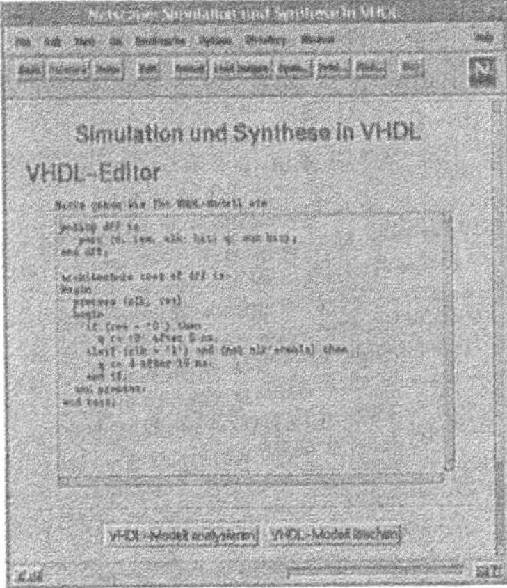

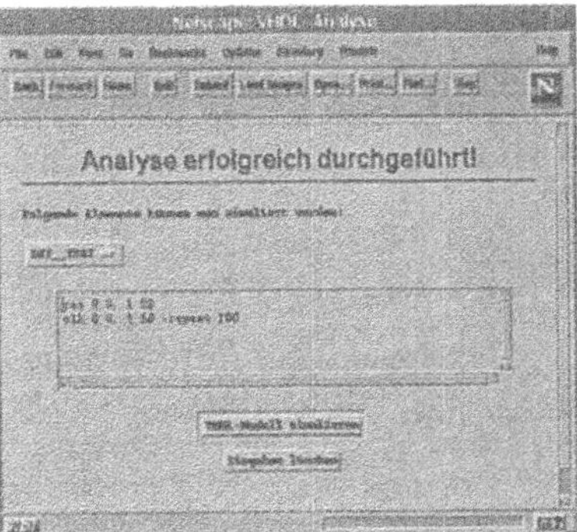

Figure 3: Integrated simulation tools for exercises

3.3 SYSTEMS WITH SEPARATE TOOLS

In the last category the learning system is used together with separate simulation and synthesis tools. This category is subdivided in on-line and off-line systems.
The on-line systems work similar to the above mentioned browser-based tools, with the difference, that the user interface of the EDA tools is displayed on the users machine in a separate window (e.g. VHDL-Online [5]). The advantage of this approach is, that the student or engineer works directly with the tool, which is also used later at the university or company. So the user trains early the special environment of the tool. The disadvantage especially for beginners is, that besides learning the complex language VHDL, the philosophy of the tool and the operation of the tool need to be understood.
Off-line systems include tools which are able to run on the local hosts, e.g. our VHDL learning system HypermediaVHDL [3] includes a simulator to do the exercises. It combines some characteristics from the second and this category: The integrated transparent start of the tools within the system and the display of the tools (editor, compiler, simulator, graphical output) in separate windows. The use of a simulator on the local system is possible, because it is a self-developed system especially for the course. The integration of a commercial VHDL system is also possible, but it is necessary to obtain licenses, which is usually not possible for university labs. HypermediaVHDL is also available on-line.
The System VHDL-Online is also available on CD-ROM for a stand-alone use, but this version is lacking the simulation and synthesis possibilities for interactive exercises.

4 Conclusion and outlook

We have shown different approaches to VHDL learning systems. On one hand there are stand-alone systems with or without user defined exercises which allow to do some VHDL lectures were ever you want. Here you concentrate on the VHDL basics. On the other hand are the on-line systems with remote compute servers which are able to simulate and synthesize even complex models using special EDA tools.
The increasing communication bandwidth even for mobile networks and the availability of EDA software on Win95/Win NT will reduce the gap between the two approaches. In the near future the CBT systems will be available at any place at the time, the user needs them.

5 References

[1] N.N.: *IEEE Standard VHDL Language Reference Manual 1076-1993*, (IEEE, New York, 1993)

[2] N.N.: *Doulos VHDL PaceMaker Users Guide*, Doulos, 1997

[3] H. Dicken, M. Koch and D. Tavangarian: *The HypermediaVHDL Learning System - Description and First Experiences*, 1st European Workshop on Microelectronic Education, Grenoble,1996, http://www.tec.informatik.uni-rostock.de/hypervhd

[4] N.N.: *Esperan MasterClass Users Guide*, Esperan, 1997

[5] N.N.: VHDL-Online, University Erlangen-Nürnberg, http://www.vhdl-online.de

THE ASTEP EDUCATIONAL MULTIMEDIA FRAMEWORK

P. FOULK[1], M. DESMULLIEZ[1], L. MACKINNON[1] and M. FERREIRA[2]
[1] *Heriot-Watt University - Edinburgh EH14 4AS - UK*
[2] *Univ. of Porto (FEUP/DEEC) – R. Bragas - Porto - PORTUGAL*

1 Abstract

This paper addresses the use of multimedia frameworks for the delivery, support and assessment of trainees for process-based high-technology companies. This framework is the basis of an European telematics network that is being developed in the ASTEP project for semiconductor fabrication, microsystem manufacturing and high-level design and test activities. ASTEP (*Advanced Software for Teaching and Evaluation of Processes*, project reference number MM1001) is a project being carried out in the framework of the Commission of the European Community Action on Educational Multimedia and will deliver a set of multimedia course modules designed for the semiconductor industry. The ASTEP consortium combines companies from this industry area and universities, with the final objective of improving the qualification of the skilled workforce required for a competitive semiconductor industry in Europe. The ASTEP strategies to achieve success, namely in what concerns the exploitation of results, the technical and pedagogical approaches and the mechanisms set up to assure constant update of course contents, are presented as well.

2 Background and rationale of the ASTEP project

Distance learning technology is now a powerful alternative to traditional classroom education, in spite of the identified disadvantages of current technologies for this purpose [1,2]. The major benefits assuring a successful future for remote course delivery are the following:

- Any number of trainees can be accommodated, allowing the organisation of training programmes with little dependence on the expected size of the target audience
- Training schedules can be easily adapted to each trainee's occupations and to their diverse needs to improve professional skills
- Access to information and interaction with lecturers / other trainees is much simpler

The range of tools available for online course development and delivery is however a complex task, requiring a detailed comparison among the alternatives available [3,4].
As a combined effort between industry and academia, the ASTEP project rationale derives from the potential of interactive multimedia telematics networks to effectively provide solutions for this and other similar needs.

T.J. Mouthaan and C. Salm (eds.), Microelectronics Education, 213-216.

3 The ASTEP project

The ASTEP project will be described in this section, starting with the project objectives and proceeding to the development model underlying the R&D work.

3.1 ASTEP OBJECTIVES AND CONSORTIUM

The purpose of the ASTEP project is to create a multimedia educational platform and a European telematics network for the delivery, tutoring and assessment of trainees employed in process-based high-technology companies. To prove the capabilities and benefits of this platform and network, the following sets of courses are being designed:

- Motorola and SensoNor semiconductor processes
- Generic semiconductor process
- Test and High Level Design

At the end of the project, the trainees will be able:

- To learn the manufacturing process and characterisation techniques at each step of the process
- To diagnose and evaluate any deviations of the process from the nominal one
- To share peers' experience and the latest technological breakthroughs through the telematics network
- To be able to acquire a Nationally recognised diploma through tutor support and assessment modules of the learned material

The objectives given above emphasise the immediate requirements of high-technology companies and the acute needs for Europe-wide accreditation procedures within a short timescale. In order to fully meet these objectives, the ASTEP consortium comprises the following members:

- The companies Motorola UK and Motorola France (MOT) for the development, testing and exploitation of the proprietary process package
- The company SensoNor Norway for the development, testing and exploitation of the proprietary process package
- The European Teaching Organisation of the company Applied Materials (AMAT) Germany for the development and testing of the generic process package
- The Colleges Buskerud Norway (HIBU) and West Lothian College (WLC) Scotland for the development, testing and exploitation of the generic process package
- The University of Porto (FEUP) combined with other Portuguese Universities, for the development, validation and exploitation of the Test and High Level Design package
- Heriot-Watt University (HWU) expertise in computer-based learning, project co-ordination and management

3.2 THE ASTEP DEVELOPMENT MODEL

The consortium encompasses enough diversity in terms of countries (France, Germany, UK, Portugal), type of partners (Industry, College, University, National Organisation) and educational abilities of the trainees (operator, technician, engineer, researcher) to demonstrate fully the pedagogical and technical capabilities of the project outcomes.

Fundamental to the nature of this project is the intention to produce a generic ASTEP framework which could be used to support training of staff for any process-based industry. As such, this means that the actual development of an instantiation of ASTEP for the semiconductor manufacturing industry, for example, provides a demonstration of the capability of the general framework to be customised to a specific industrial domain. Within this context, the Industrial, and to some extent the Learning Institution, partners within the consortium, will evaluate the effectiveness of this customisation process, as their focus throughout the project will be to consider ASTEP in terms of the semiconductor industry. Therefore, each individual component of the framework will be developed and tested as a general component by the multimedia and telematic expert partners within the consortium, and will then be customised for the semiconductor industry and evaluated and assessed by those partners whose focus is on that particular domain.

Once a generic model of ASTEP has been developed, a standard commercial alpha-beta test model will be adopted. Alpha testing will be carried out at Industrial and Academic partner sites with test data, related but not necessarily specific to the domain, instantiated into the framework.

The framework will be evaluated for usability, consistency, speed, and the efficiency of the delivery of material, specifically multimedia assets across telematic links. As part of this process, research staff from the multimedia and telematics expert partners will be based in the companies and learning institutions, to support the establishment of the infrastructure for ASTEP, to provide a basis for technology transfer and to aid in the capture and collation of evaluation results.

Without pre-empting on the dynamic links to be established for a full demonstration of the product capabilities, one could envisage the following scenario in order to demonstrate a typical course delivery illustrated in figure 1.

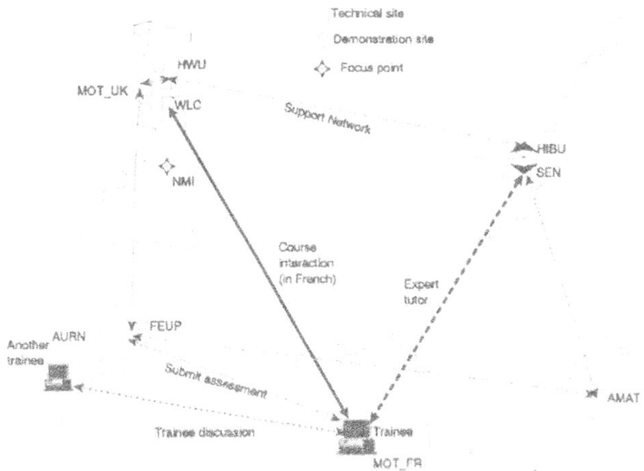

Figure 1: A typical course delivery scenario.

A trainee in Toulouse wishes to learn the package devoted to the generic process manufacturing at WLC. SensoNor happens to deliver a colloquium on microsystems which is being videoconferenced to the partners of the network. The trainee follows this conference and interacts with the expert tutor from Norway. This conference is also followed by employees in AMAT. After the conference, the trainee comes back to his module which is being supervised at the moment by some tutors at WLC. A course interaction takes place with the tutor on specific points of the module. The trainee has the freedom to consult also others trainees (such as those in Portugal) who happen to study the same course. At the end of the module, an assessment is completed by the trainee and

sent to FEUP which is assessing the modules. HWU ensures that the management of the network is adequate with its partners HIBU and MOT_UK for example.
Beta-test results will be incorporated into a final, product release of ASTEP as a generic framework, and an industry-specific release of ASTEP for the semiconductor manufacturing industry. User group newsgroup, mailing-list, or other appropriate telematic structures will be developed to provide ongoing support and feedback structures relative to the further development, maintenance and exploitation of ASTEP. This structure will be distinct from the support framework provided within ASTEP itself, for users in a particular domain. As part of the exploitation plan for the project, appropriate feedback response times, formats and actions will be identified and responsibilities assigned to partners within the consortium.

4 Conclusion

The ASTEP project was set up in response to the difficulties experienced by traditional engineering education in the area of semiconductor manufacturing. The need to cope with the exceptionally fast pace at which the technology evolves in this area was another reason behind ASTEP, since a permanent update of the technical skills of employees in process-based high-technology companies is vital in terms of market competitiveness.
By delivering flexible training programmes through an educational multimedia telematics network, the project helps to improve and maintain European competitiveness in the global semiconductor industry and allows the creation of a pool of trainees geographically dispersed, which can be identified easily and employed by potential new companies. It enables the rapid propagation, by experts, of technical breakthroughs and state of the art reviews, while at the same time creating a critical mass of trainees, which is particularly effective for the dissemination of solutions on internal technical issues. For SMEs and international companies alike, this type of course delivery is also very effective for the training of rapidly expanding activities critical to the SMEs future. Another important issue is that the proposal aims at developing qualifications of trainees which are officially recognised, enhancing the mobility of labour within Europe and facilitating the quality audit process of trans-national companies. Finally, and in economically deprived areas which are often attractive to multi-national companies for financial reasons, the delivery of courses by telematics facilitates the retraining of unemployed workers in suitable high technology skills.

5 References

[1] The University of Manitoba, *Advantages and Disadvantages of Web Based Instruction.*
[http://www.umanitoba.ca/ip/tools/courseware/pros.html]

[2] Robin Mason, *The Globalisation of Education*, 1997.
[http://www-iet.open.ac.uk/staff/robinm/GlobalEdu.html]

[3] Herb Bethoney, "Computer-based training on the web," PCWeek, August 1998.
[http://www8.zdnet.com/pcweek/reviews/0818/18ibt.html]

[4] Bruce Landon, *Online Educational Delivery Applications: A Web Tool for Comparative Analysis*, September 1997.
[http://www.ctt.bc.ca/landonline/index.html]

MICROWIND : INTRODUCING MICROELECTRONICS DESIGN ON PC

Etienne SICARD, CHEN Xi

INSA, Department of Electrical & Computer Engineering
Av de Rangueil 31077 TOULOUSE Cedex 4 - France
Fax: +33 561 55 98 00 -
e-mail: etienne@dge.insa-tlse.fr
web page : http ://www.insa-tlse.fr/~etienne

1 Abstract

This paper presents a PC based software running on PC/Windows-95 dedicated to the training in sub-micron CMOS VLSI design. The software consists in a HDL-based schematic editor with on-line simulator, a layout editor with HDL-compiler and sophisticated on-line analog simulation. Furthermore, the tool includes a set of tutorials covering various aspects of integrated circuit design : tutorial on CMOS process, on MOS devices with emphasis on electron-level behavior, models and analog basic cells.

2 Schematic Editor DSCH

The schematic editor handles both conventional pattern-based logic simulation and intuitive on-screen mouse-driven. A library of standard symbols with HDL model description is also provided. On the screen reported in Figure 1, the simulation of a 4 bit adder built from hierarchical adder gates is shown. On a click on the Adder symbol, the symbol properties appear : HDL description, list of pins, properties attached to the pins and various other options.

T.J. Mouthaan and C. Salm (eds.), Microelectronics Education, 217-220.

The student can describe the behavioral description of his component and verify its functionality by direct simulation. A library of standard symbols with HDL description in terms of basic logic primitives are also provided. The logic simulator performs a translation of the HDL description into an internal code to speed up the simulator efficiency. Figure 2 reports the Verilog description of the 4-bit adder generated by the schematic editor.

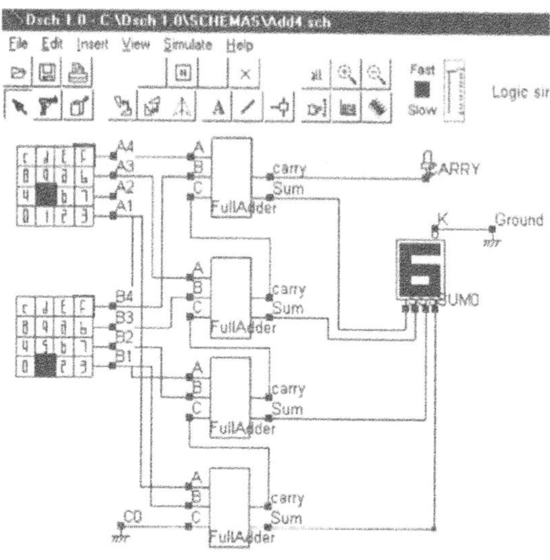

Fig.1 : Simulation of a four bit adder

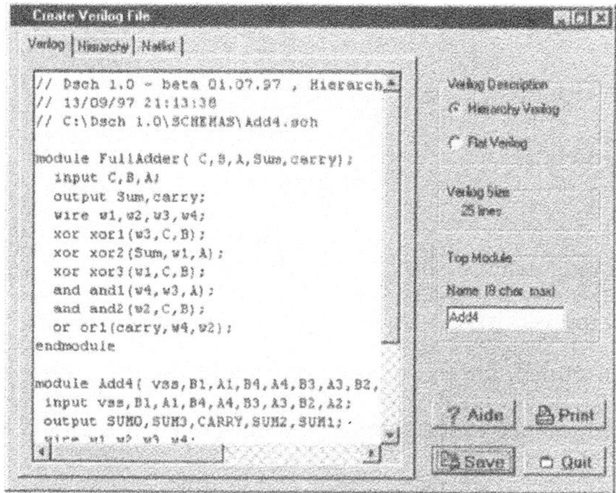

Fig.2 : Verilog description of the Adder

3 Layout Editor

The layout editor features full sub-micron layout design, simulation and verification. Up to 2000 device layouts can be handled and simulated easily in a Pentium 8Mb PC. A new significant tool is the possibility to compile HDL logic description issued from the schematic editor directly into layout. The 4-bit adder described in Fig.1 is compiled into layout with a result reported in Fig. 3.

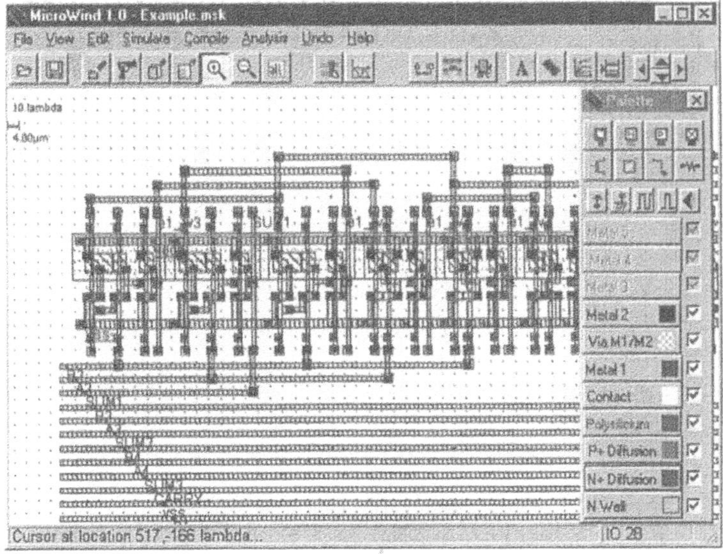

Fig.3: Layout synthesis of the 4-bit Adder from its Verilog description

The cells are placed in a linear manner, with horizontal VDD and VSS power planes, n-channel MOS devices on the bottom and p-channel MOS devices on the top. The internal cell routing is built on the top of the active area and the bloc routing is generated on the bottom. The metal1 layer acts horizontally and the metal2 layer is used vertically. The width and length or the n-channel and p-channel MOS network is user-specified. The advantages of the cell compiling are its adaptability to any design rule set, any criterion, so that no predefined layout library is required. The cell compiler handles single MOS, basic gates, complex gates, and edge-sensitive registers. All other cells must be constructed using these primitives.

The analog simulation of the circuit is invoked at the press of one single key (Figure 4). To realize this task, an electrical extraction of the circuit is automatically performed and the analog simulator produces voltage and current curves by solving numerically in time domain the network equations, basically currents flowing through MOS, parasitic capacitance and resistance.

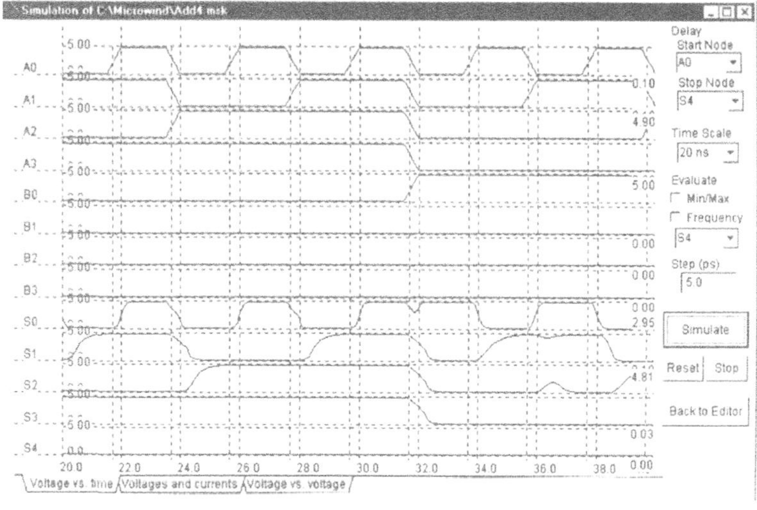

Fig.4 : Analog simulation of the 4-bit Adder.

A set of educational tools are also included in the layout editor : 2D vertical aspect of the process, tutorial on MOS device parameters, measurements performed on real test chips, and parametric analysis.

4 Conclusion

The software has been in use for one year in around 50 universities. The student feedback was globally very satisfactory. Further development will include a tutorial on MOS model with an HDLA interpreter and an electron-level monte-carlo simulation.

5 References

[1] E. Sicard, Chen Xi «Microwind, Dsch, an Introduction to micro-electronics », INSA Edition, Toulouse, France, 1998, ISBN 2-87649-017-X

EXPERIENCES OF COMPUTER-AIDED-LEARNING IN DEGREE-LEVEL ELECTRONICS

J.N. COLEMAN, D.J. KINNIMENT, G. RUSSELL,
*A.M. KOELMANS,
*Dept Electrical and Electronic Engineering; *Dept Computer Science,*
The University,
Newcastle upon Tyne,
NE1 7RU,
U.K..

1 Introduction

Over the last few years, a collaborative project within a group of British universities has developed a suite of computer-aided-learning material covering a large proportion of the earlier years of their Electrical Engineering courses. This courseware has subsequently been made publicly available under the Electronic Design Education Consortium (EDEC), and many parts of it are now in use throughout the U.K.. Although this new material is becoming familiar to a wide audience, there has so far been little analysis of the impact of its adoption on teaching techniques, course structures and design, and student learning experiences.

At Newcastle upon Tyne the first courseware module was put into use some five years ago. Since then, a great many more modules have come into use, and it is the purpose of this paper to describe some preliminary observations on the ways in which large-scale adoption of computer-aided-teaching has developed. It is not possible at this stage to present any scientific study or conclusive evidence about the efficacy or otherwise of these techniques, but we can describe avenues for further investigation which are now opening.

2 The Courseware

Typically a degree course consists of a number of lecture courses lasting about one term. The selection of such courses, and their detailed content, varies very widely between different universities. There is far too much variability to make a common suite of courseware feasible. EDEC sought to overcome this problem by adopting a strategy of very fine-grained modularity. The material is divided into a large number of very small modules, each the equivalent of only a few hours of lectures. Although the modules group into sets with approximately the same coverage as a traditional one-term course,

T.J. Mouthaan and C. Salm (eds.), Microelectronics Education, 221-224.

the lecturer has the option of using all or any of these individual units. If a lecturer is satisfied that a module adequately covers part of his lecture material, he may wish to use it; otherwise he will continue lecturing as before.

The outcomes of this approach will clearly lie on a continuum between fully or almost-fully lecture-based courses, and fully computerised courses. In the following section we describe four examples which cover this range, and on the reactions of staff and students to them.

3 Curricula

Each of the following subjects is taught in a course lasting about a term in either the Electrical Engineering or Computer Science Department at Newcastle upon Tyne. Each covers the sub-areas shown, which are taught either by lectures (LECT) or by courseware (CW). In some cases the CW is used simply for revision, or as an introduction when lectures are then used to expand on the detail. In others the lectures are eliminated and the CW is supported directly by practicals, tutorials, or notes (which include written exercises). In the latter case, the time spent by the students on a CW module is typically less than half of the time previously spent attending the corresponding lectures. Of course, this difference only becomes significant when a large number of CW modules are used together.

Digital Electronics	
Design for Test	LECT
Fault simulation, ATPG, ATE	LECT, optional CW for revision
Simulation	LECT
High Level Synthesis	LECT and CW simultaneously
System Design	LECT
Machine and System Architecture	
Number Systems	CW intro then LECT
Digital Logic	CW intro then LECT
Computer Architecture	CW intro then LECT
68000 Assembly Language 1 (Introductory)	CW intro then LECT
68000 Assembly Language 2 (Intermediate)	CW intro then LECT
Microprocessor Systems Design	
Review of Computer Architecture	CW (review of LECT)
Review of 68000 Assembly Language 1	CW (review of LECT)
Review of 68000 Assembly Language 2	CW (review of LECT)
Microprocessor Systems	CW with NOTES
Digital Interfacing	LECT
Memory	LECT

I/O Controllers and Programming	LECT
Design Project	PRACTICALS
Computer Engineering	
Number Systems	CW with NOTES
Computer Architecture	CW
68000 Assembly Language 1 (Introductory)	CW with PRACTICALS
68000 Assembly Language 2 (Intermediate)	CW with PRACTICALS
Design Project	PRACTICALS

4 Experience of Use

In the case of *Digital Electronics*, the CW material was used in two ways. First, it was used as an optional backup to one part of a conventional lecture-based course. Student reaction to this was poor; there was very little take-up and the lecturer did not report any noticeable difference in the effectiveness of the teaching, or any significant impact on the design or structure of the course. It appeared that students felt reluctant to commit time to this activity unless it were more strictly integrated into their work schedule. However, another CW module was used at the same time as the corresponding lectures, in a novel approach in which the students could try the material as the lecturer spoke. Student response to this was highly positive.

More direct application of CW material was evident in *Machine and System Architecture*, where several modules were used as introductions to the lectures, which repeated and expanded on the CW material. Printed notes were available, but these were primarily a re-statement of the basic factual material. We are forming the opinion that this approach was liked by the students, but it was found to be important to stress that attendance was compulsory, and to take a register at the CW sessions.

In *Microprocessor System Design*, CW was used primarily for review purposes, (the students having covered this material the previous year). No printed backup was therefore thought to be necessary, or requested by the students. Indeed they did seem to appreciate being able to move through this revision material somewhat faster than they would if they had attended lectures. However, one CW module was used to present new material as a direct replacement for lecturing. This module was accompanied by printed notes with suggestions for further work. The remainder of this course was taught in the conventional way, and was reinforced by a large-scale practical exercise which brought together most of the strands of the teaching.

In *Computer Engineering*, the CW was used to the fullest extent, entirely replacing the lecturing. In almost all cases the CW was supported by additional printed notes, further work, or practical exercises. The students expressed a preference for at least a few lectures during the course, so one or two new lectures on supplementary topics were introduced. The difference in overall delivery time became significant, and was used to provide more practical sessions distributed throughout the course. During these sessions, the lecturer could also interact at a personal tutorial level with individual students. Analysis of the relevant assessment marks has shown little change over the last few years as the CW modules have gradually replaced the lectures, although this process has only just reached completion and it is too soon to say what will be the long-term effect.

5 Discussion

At the moment it is far too soon to draw any conclusion as to the benefits that computerised teaching might bring, although few points do stand out already. One is that the use of CW material must be tightly integrated into the overall course structure, preferably with prescribed attendance requirements, and not merely left as an optional extra. It is also evident that the students appreciate the additional variety that this type of teaching brings, and that if the lecturing is substantially reduced then additional types of activity such as practical work, tutorials, or written work should be brought in to compensate. There is clearly a wide variety of different ways in which the CW can be used, ranging from an 'animated lecture' to use as the primary teaching method.

These different approaches will be studied carefully over the next few years. Points to be scrutinised will include trends in examination marks, performance of students in later years as an indicator of long-term retention, measures of delivery time, lecturers' and students' observations, and timetabling and resource requirements. A new project has recently been announced in which some members of EDEC will collaborate to investigate these issues further.

Session P6

DESIGN INNOVATIONS

INTRODUCTION TO DIGITAL INTEGRATED CIRCUIT DESIGN

M. JEZEQUEL, P. ADDE and G. GRATON
Département Electronique
ENST de Bretagne
BP 832, 29285 BREST CEDEX, FRANCE

1 Abstract

This paper presents a practical course followed by 2nd year students at the Ecole Nationale Supérieure des Télécommunications de Bretagne. During this practical course students have the opportunity to design an integrated circuit using the Ping-Pong technique. Field Programmable Gate Array technology is used in order to allow the circuit to be tested with real signals.

2 Context

ENST Bretagne (Ecole Nationale Supérieure des Télécommunications de Bretagne) is a French telecommunications engineering school where entrance to the first year as an engineering student is possible by passing a joint competitive examination common to some other "grandes Ecoles" (graduate engineering schools). The engineering curriculum at ENST de Bretagne is in three parts. The first three semesters form the common core and give an overview of the telecommunications world. The fourth semester gives a first approach to engineering methods. In the third year, the option in the fifth semester represents their first specialisation, then the final year project (6th semester) gives the students a first taste of professional life.
During the common core the students follow a practical course called the "experimental project". It places them in a teamwork situation with problems of methods and experimentation for a total of 24 hours. The students choose one subject from the eight proposed.
One of the proposed subjects is entitled "digital integrated circuit design". In this project, students have to design two circuits called supervisors. They implement the Ping-Pong technique.

3 The Ping-Pong Technique

The Ping-Pong technique (or time division duplex) finds applications in any bi-directional digital communication link where time division multiplexing is better than

T.J. Mouthaan and C. Salm (eds.), Microelectronics Education, 227-230.

frequency multiplexing or echo cancelling (for example: off-line telephone, telematic links for short distances, ...).
Figure 1 shows a system where the Ping-Pong technique is used. Station A and station B are linked by a single channel. This channel allows transmission from A to B or from B to A but not in full duplex. The Ping-Pong technique gives a perfect illusion of full duplex mode transmission on this kind of channel.

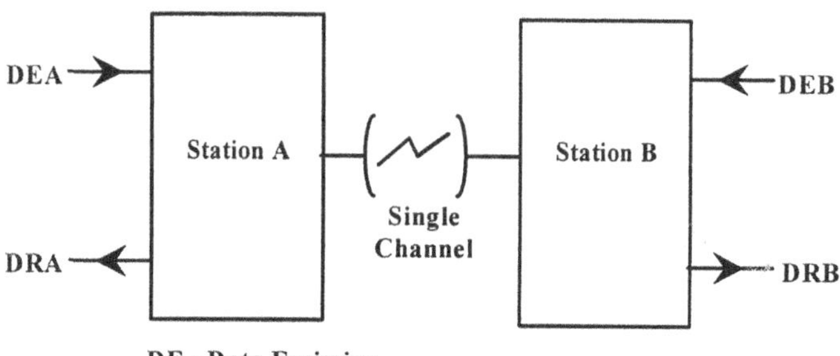

Figure 1: Digital communication using the Ping-Pong technique

Circuits designed by students are the supervisors of stations A and B. As is shown in figure 2, the supervisor of station A changes the continuous emission data DEA (Data Emission for station A) to packet emission PEA (Packet Emission for station A). After a propagation time, the supervisor of station B receives these packets then converts them into continuous data DRB (Data Reception for station B). The transmission from station B to station A follows the same process.

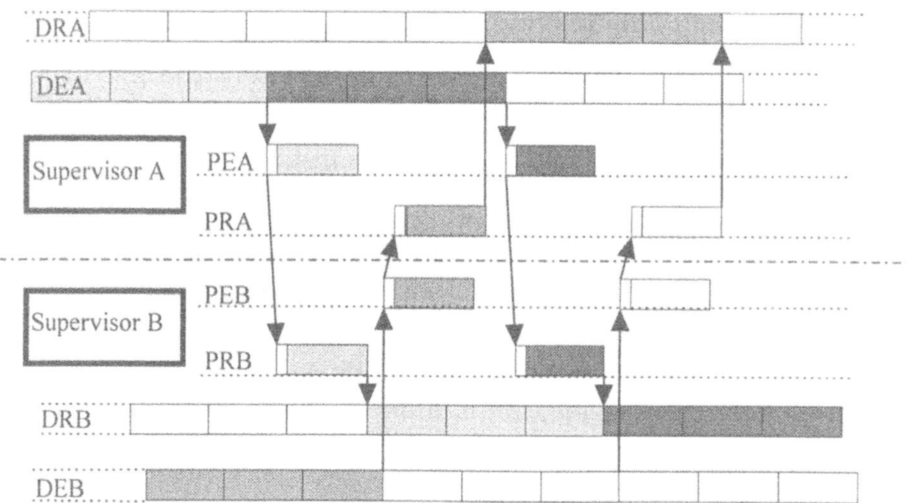

Figure 2: Packet exchange

As it is a synchronous link the clocks of station A and station B have to be synchronised. So the two supervisors cannot be identical because one of the stations must provide the reference clock.

The ratio between the packet data rate and the continuous data rate is equal to three. Packets are built from useful information (16 bytes) to which control and signalling bytes are added. The two first bytes are used for clock synchronisation, the third for packet synchronisation and the last for error detection. The fourth byte allows information exchange in order to make the link management protocol possible.

4 Student assignment

The project starts with a lecture which presents the "V method" [1], which is the design method used during the project. The Ping-Pong technique is presented too. At the end of this lecture the group of students is split into two teams of 10 or 12 people. Each team is responsible for one station (A or B) with the help of one lecturer.

First of all the lecturer presents the system requirements of the station to the team. Then he focuses his presentation on the supervisor which is the circuit to be designed. Students do not actually have to design the circuit completely because there is not enough time to make it in 24 hours. During the presentation of the general architecture of the circuit the lecturer shows the part to be designed.

In fact a first version of the two supervisors was designed by third year students during a practical course [1]. These circuits are made up of eight main blocks: three for the emission (scheduler, memory, packet construction), three for the reception (scheduler, memory, packet decoder) a protocol manager and a clock generator. Blocks implementing memories are not designed by students because they lack any pedagogical interest. The clock generator block includes a DPLL (Digital Phase-Lock Loop) [2] used for clock recovery which is too complex to be designed by students during this project. They have therefore to design five blocks: the schedulers, the packet construction, the packet decoder and the protocol manager.

The schedulers are built around two counters: a Gray counter and a binary counter. Students have to synthesise the two counters and certain outputs following block requirements.

The packet construction block is built around multiplexers and an LRC (Longitudinal Redundancy Check) generator. This generator gives a byte where the parity of each bit is respected for the whole message.

The packet decoder block is the reverse of the packet construction block. An LRC generator is included in it in order to make possible the detection of transmission errors.

The protocol manager block is the implementation of a state machine. The state chart is given to the students. There are five states and seven inputs.

The lecturer concludes his presentation by showing the set of elementary logical functions (D flip-flop, Nand, Nor, ...) that students will use for the synthesis. Then they share the work out among themselves. From that time on, they work in five groups of two or three people. Each group is responsible for one block. They start by manually synthesising the block which they are in charge of.

At the end of the synthesis period, students have to propose functional test patterns for the block. Next they use COMPASS CAD tools in order to capture the schematic of their block. Then they simulate it using their functional test patterns in order to check their work. If design errors are detected they come back to the synthesis or to the schematic capture.

After this step the team receives a design capture of the circuit where the five blocks to be designed have been erased and the students have to insert their design in it. Then they simulate the global design with reference patterns given by the lecturer. This step allows the whole project to be checked. Of course, if simulation results are not in accordance with the system requirements then students have to come back to the block design.
When simulation results are good the circuit is placed and routed on an FPGA (Field Programmable Gate Array). The circuit used is an XILINX XC4010. About 70% of the CLBs (Configurable Logic Block) are taken up by the Ping-Pong supervisor. This technology allows circuit testing during the experimental project.

5 Circuit testing

In order to make tests with real signals possible, a board was designed. This board can be used as station A or station B. It simulates a single channel for transmission and generates all the signals needed for the tests. It is made up of:

- an FPGA XILINX XC3020 associated with a quartz which generates clocks and pseudo-random sequences used for testing the transmission link,
- analogue components making possible simulation of a single channel transmission link,
- switches and LEDs used for programmability and visualisation of certain signals,
- the supervisor (an FPGA XILINX XC4010) which can be loaded from a personal computer.

So, when students are quick enough they can test their own circuit; otherwise they can see another circuit in use.

6 Conclusion

A first experience of this practical course has shown that students are satisfied with this project. Moreover, it has motivated many of them to choose the option focusing on Integrated Circuit Design during their third year.

7 References

[1] C. Douillard, P. Ferry, P. Adde, M. Jézéquel, "The introduction of VHDL in digital design practical courses", 1st European Workshop on Microelectronics Education, Villard de Lans, France, pp. 189-192, February 1996.

[2] J.R. Cessna, "Digital phase-locked loop with sequential loop filters: A case for coarse quantization", Proc. Int. Telemetering Conf., vol. VIII, pp.136-148, Oct. 1972.

DESIGN OF A MULTI-FPGA SYSTEM FOR RAPID PROTOTYPING EXPERIMENTATION

J.-D. LEGAT, J.-P. DAVID
Université Catholique de Louvain
Microelectronics Laboratory
Place du Levant, 3
B-1348 Louvain-la-Neuve
Belgium
Email : Legat@dice.ucl.ac.be

1 Introduction

FPGA allows today developing fast prototyping. The complexity of the system to be implemented is however limited by the number of gates available in the FPGA. This barrier can be dropped by considering multi-FPGA architectures [1,2]. The application implementation is then split into different circuits. Many problems such as system partitioning, simulation and synthesis are associated with multi-FPGA systems. This paper will present the use of a multi-FPGA-based PCI board by students to experiment rapid prototyping of digital signal processing algorithms.

2 Reconfiguable Coprocessor Architecure

The system architecture is shown in figure 1. It is based on an enhanced mesh topology which makes block control and data transfer with the host computer easier [3]. This architecture implements a linear mesh topology [4] coupled with a global and limited crossbar which gives each FPGA a direct connection to each part of the system including the PCI interface. Instead of using a large number of medium size FPGA with crossbar for connecting 2 adjacent circuits, a solution based on high density FPGA has been adopted because it allows to reduce external connections and to avoid additional crossbars. These connections can be hardwired without strong penalty during the place&route process.

The Flex 10K100-3DX from Altera has been selected due to its high capacity (4992 logic elements (a 4-input LUT and a 1-bit register) and its internal RAM (24 Kbits) that can be easily reconfigured and used as large look-up tables to implement DSP functions. Additional external fast static RAM has been provided. This memory is organized in 32-bit words where each byte can be disabled. 2 Mbytes are associated with each FPGA. The whole system is equivalent to 400Kgates with 8Mbytes external

T.J. Mouthaan and C. Salm (eds.), Microelectronics Education, 231-234.

SRAM on a single PCI board. The PCI interface is based on a small dedicated FPGA which moreover implements the crossbar and Flex 10k configuration, the JTAG and the system clock signals. This device is automatically configured by a removable EPROM at power-on or at hardware reset.

This system acts as a reconfigurable coprocessor working in parallel with the host processor.

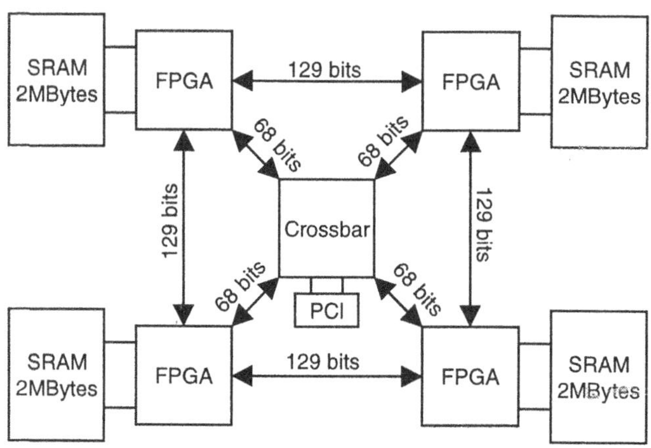

Figure 1: The enhanced mesh topology

3 Rapid Prototyping Experimentation

3.1 FIR IMPLEMENTATION

Preliminary experiments of rapid prototyping have been performed. The first one concerns the implementation of FIR for audio processing [5]. In this work, the challenge consists in implementing efficiently the MAC operation (multiply-accumulates). A 2048-coefficient FIR at 48 KHz sampling rate for professional audio has been successfully tested.

The coefficients are preliminary stored in external RAM whereas the data are serialized in the internal RAM. They go from one processing unit to the other. In this way, the system can switch instantaneously from one set of coefficients to another by simply addressing another block of external RAM. The large capacity of external RAM allows storing up to 1024 sets of 4096 16-bit coefficients. This unusual feature, which cannot be implemented in fully parallel distributed arithmetic [6] (where the filter coefficients are fixed) offers a lot of new possibilities for real-time adaptive signal processing. The coprocessor can accept samples at 48Khz rate for professional audio filtering. This application uses 2/3 of the internal RAM resources of the FPGA but only 50% of programmable logic.

3.2 SUBBAND TRANSFORMATION

The second experiment is related to the subband coding. Each FPGA is made of basic logic cells (a small 4-input look-up table with latch) and the complexity of the multiresolution implementation can be expressed in terms of logic cells. Taking into account that the coefficients of the multiplication are constant, the 16-bit by 8-bit multiplication true tables can be stored in look-up tables. A whole lattice filter stage (L=8, N=3, 8 Multiply, 8 Add) only needs 924 logic cells which allows to execute in real-time multi-stage multiresolution transform on CCIR601 television signal [7].

Thanks to pipelining and logic optimisation, a clock frequency of 40 MHz has been obtained which gives a computation power of 640 MOPS for this basic block. Due to large number of logic cells available (up to 20k), the lattice block filter can be easily duplicated in order to obtain the desired requirements. Without considering control and data transfer, a theoretical computation power peak of 11 GOPS can be achieved (20k/1104*640 MOPS). External RAMs are used to store input image, intermediate results and subbands.

3.3 DATA-FLOW ORIENTED CODESIGN

Most reconfigurable systems contain external RAM for global storage. For some years, FPGA manufactures have offered new possibilities of logic synthesis by adding small blocks of RAM embedded with reconfigurable logic. These characteristics offer new possibilities for hardware/software co-design in terms of memory management. As for standard processor architectures that include one or two cache levels, FPGA based systems can also take into account the spatial and temporal locality properties for data.

Instead of using only large external RAM which have intrinsic long delays, we have proposed to use internal embedded RAM as dedicated caches [8]. Moreover, thanks to the large amount of logic available, the concept is extended and the general architecture can be divided into two main parts : the tasks which includes the processing units and local RAM, and a transfer controller which handles all transfers between external RAM and tasks. Tasks are thus waiting for data. These are processed as soon as they are available. This is the property of dataflow architecture.

This method has been tested on a C program that computes a 3x3 convolution on a 256x256 pixel image. The program requires 590K read operations and 65K write operations applied to global data. Thanks to local RAM caches, this leads to 197K read operations and 66K write operations. On the multi-FPGA architecture running at 33Mhz clock, we obtain a time improvement from 20ms to 8 ms.

4 Conclusions

These preliminary experiments have allowed to validate this multi-FPGA architecture and have pointed out the interest to develop specific tools to help the designer. In the next academic year, students will intensively used this rapid prototyping system within the framework of final works or projects related to courses. Examples of projects which will be carried out by students are the implementation of a full CDMA system, an IEEE 1394 serial bus interface, an USB serial bus interface, a complete DV (Digital Video) decoder. Research will be performed in the following areas : rapid prototyping methodology, system partitioning, VHDL synthesis on FPGA and embedded systems design.

Figure 2: Board photograph

5 References

[1] J. Vuillemin, P. Bertin, D. Roncin, M. Shand, H. Touati, and P. Boucard, "Programmable Active Memories: Reconfigurable Systems Come of Age" in IEEE Transactions on VLSI Systems, Vol. 4, No 1, March 1996

[2] S. Hauck, G. Borriello, C. Ebeling. "Springbok: A Rapid-Prototyping System for Board-Level Design", ACM/SIGDA 2nd International Workshop on Field-Programmable Gate Arrays, Berkeley, February, 1994.

[3] J.P. David, J.D. Legat. "A 400Kgates 8 Mbytes SRAM multi-FPGA PCI system ". Int. workshop on logic and architecture synthesis, Grenoble, 16-18 dec 1997, pp 113-117.

[4] S. Hauck, G. Borriello, and C.Ebeling, "Mesh routing topologies for multi-FPGA systems, in Proc. Int. Conf. Computer Design, 1994, pp. 170-177.

[5] J.D. Legat, J.P. David. "A multi-FPGA based coprocessor for digital signal processing". IEEE Benelux Signal Processing Systems Symposium, Leuven, March 1998.

[6] Bruce Newgard "Seminar on signal processing with Xilinx FPGAs", Xilinx , 1996

[7] J.D. Legat, J.P. David, P. Desneux. "Programmable architectures for subband coding : FPGA-based systems bersus dedicated VLSI chip". CESA'98 Multiconference, Computational Engineering in Systems Applications, April 1998, Tunisia.

[8] J.P. David, J.D. Legat. "A data-flow oriented co-design for reconfigurable systems". 9th IEEE International Workshop on Rapid System Prototyping, Leuven, June 98.

CIRCUIT FAULT DETECTION USING INDUSTRIAL CADENCE SOFTWARE

C. BARON, P. BOURDEU D'AGUERRE, F. CAIGNET,
A. FERREIRA, J-C. GEFFROY
LESIA, INSA TOULOUSE
Complexe Scientifique de Rangueil
31077 Toulouse Cedex, FRANCE.
Email : baron@dge.insa-tlse.fr
Fax : +33 5 61 55 98 00

1 Introduction

With technology improvement and growing complexity of micro-electronic circuits, the verification of the system functionality has become one of the major current problems [2]. In the educational context of the engineering Training Pole in Micro-electronics located in Toulouse, France (PFMT), several lectures and their associated labs on electronic systems reliability have been recently established. The PFMT regroups students in their last year of studies from five engineering schools and universities. This paper presents (section 2) the main outlines of the lecture concerning on-line/off-line testing methodologies [3], [4], which are nowadays an industrial requirement. Examples of application are also described in section 3.

2 Objectives of the testing lecture and labs

The course presents a wide range of techniques used to make electronic systems reliable. It shows how to remove faults using different fault propagation or test sequence generation methods, like the IBM ATPG (Automatic Test Pattern Generation).

The objective of the labs [1] is to test four combinatory or sequential circuits using the Verilog compiler of the CADENCE software. Three selected circuits are described at basic gate level using HDL language : a full adder, a simple memory (4*4 bits) and a synchronous counter (16 bits). The last one is a sequential system described at state-

T.J. Mouthaan and C. Salm (eds.), Microelectronics Education, 235-238.

level. Students must determine minimal test sequences from the schematic diagram of the circuits and apply them to the real circuits using the simulation tool Verifault. The two main goals of the labs are :
- to give students a practical experience of simulation tools used in the industry,
- to illustrate testing problems and demonstrate the efficiency of various test techniques.

3 Examples of application

The four circuits studied during the labs show complementary test problems :
- the four bit RAM must be tested by a minimal functional sequence detecting any stuck-at fault. The minimal sequence found by the students is estimated to 48 vectors.
- the counter comes from a real industrial case, and deals with a fault that can't be systematically revealed according to the inputs values and the counter output value.
- the fourth circuit is tested according to a functional approach at state-level.

This section details two significant examples : a simple combinatory example first, the full adder, described at the basic gate level and which can correspond to an implementation document, then a sequential example, described at state level, which can correspond to a design document.

3.1 THE FULL ADDER EXAMPLE

Figure 1 presents the gate level description of a full adder.

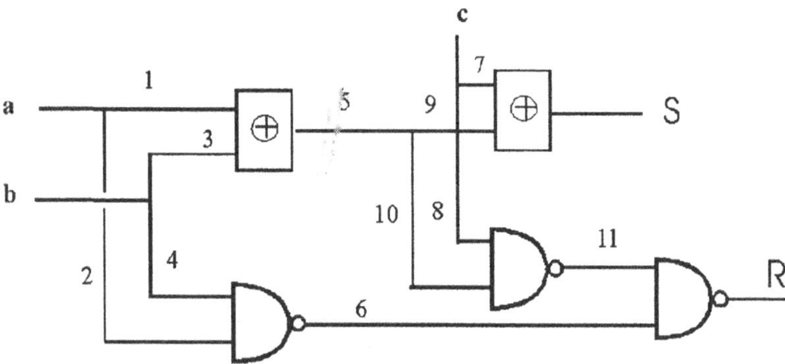

Figure 1 : Schematic description of the full adder

The goal is to study fault coverage obtained with several test sequences. The testing method is here based on the IBM D-Algorithm. The fault simulator of CADENCE is used to analyze detected and non-detected faults sequences build by students. They first have to try an intuitive test pattern with a toggle of 100% using Verilog simulator.

```
*** TOGGLE TEST SUMMARY ***
   # monitored simulated nets - 8
                   # toggled - 8
                   % toggled - 100.000000%
   # monitored port bits     - 15
               # toggled     - 15
               % toggled     - 100.000000%
148 "/home/bda/TP_FSIM/data/full_adder.run1/test.v": $finish at simulatio
41 simulation events + 22 accelerated events
CPU time: 1.7 secs to compile + 0.3 secs to link + 0.0 secs in simulation
```

Figure 2 : Summary toggle test using Verilog (Toggle=100%).

This result (fig 2) does not reveal a real fault detection, it only gives us the ratio of the activated net. By switching with Verifault-XL simulator they must complete this sequence to obtain a full fault coverage (fig 3). The final goal is to find a minimal test pattern.

```
statistics
                      Total #     Total %     Prime #     Prime %
Untestable                  0                       0
Drop_detected              30        83.3          23        88.5
Detected                    0         0.0           0         0.0
Potential                   0         0.0           0         0.0
Undetected                  6        16.7           3        11.5
All                        36                      26
end_statistics
79 simulation events + 38 accelerated events
CPU time: 1.6 secs to compile + 0.0 secs to link + 0.2 secs in simulation
End of VERIFAULT-XL 1.4   Jan 12, 1998  10:55:59
```

Figure 3 : Same Pattern using Verifault (6 undetected stuck-at faults).

3.2 THE SEQUENTIAL MACHINE EXAMPLE

Figure 4 represents the state level description of a sequential synchronous Moore machine which possesses one input ***i*** and one output ***o***.

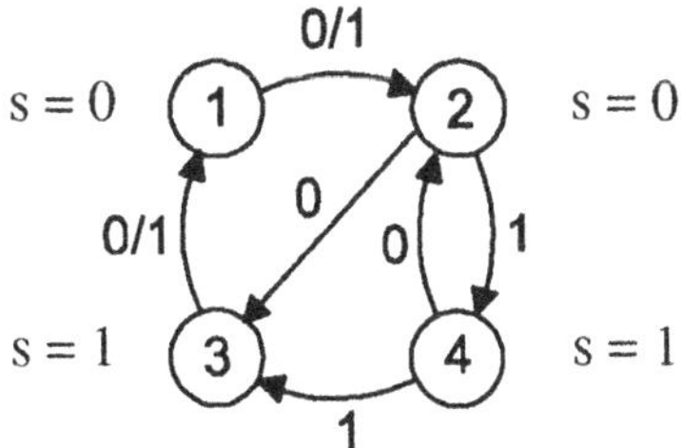

Figure 4 : Description of the sequential machine

The goal is to study the different fault coverage results obtained with two types of test sequences :
- sequences which activate each state at least one time (objective 1),
- sequences which activate each transition at least one time (objective 2).

For example :

to reach the first objective, we suggest the application of sequence 1 :

i	0	1	1	0
q	2	4	3	1
o	0	1	1	0

and obtain a fault coverage of 64%.

to reach the second objective, we suggest : the application of sequence 2 :

i	0	1	1	0	1	1	0	0	1
q	2	4	3	1	2	4	2	3	1
o	0	1	1	0	0	1	0	1	0

and obtain a fault coverage of 91,5%.

As a conclusion on these results, we can say that none of the previous approaches used allows us to obtain a 100% fault coverage. Therefore, we should add more test vectors to the previous sequences. Indeed, if we add the vectors

i	1	0
q	2	3
o	0	1

to sequence 2, we obtain a fault coverage of 100%. One must mention that some techniques, called Identification techniques [5], which are not detailed there, help to find the minimum sequence providing a 100% coverage.

4 Conclusion

We have been conducting this training for two year. The labs present a good illustration of the theory provided in the lectures by showing students the complexity of testing even simple systems. Students then realize how important and difficult the testing of real industrial systems is. During other labs, they can also manipulate on a real industrial test system (LV500 Tektronix) [6] where they compare the application of their patterns on real faulty and non-faulty circuits.

5 References

[1] http://www.aime.insa-tlse.fr/tp_fautes/

[2] B. COURTOIS « CAD and Testing of Ics and Systems, Where are we going ? » reprinted from Journal of Microelectronic systems Integration, Vol. 2, No. 3 1994.

[3] « VLSI testing » T.W.Williams editor, Advances in CAD for VLSI, Vol. No. 5, North-Holland, 1986.

[4] A. Miczo, « Digital Logic Testing and Simulation » Harper & Row Pub., New York, 1986.

[5] C. BARON « Identification of Sequential Systems » , PhD report, INSA Toulouse, nov. 1995.

[6] http ://www.aime.insa-tlse.fr/cours/

ACCURATE TIMING MODELING AND DESIGN-KIT DEVELOPMENT FOR A SUB-MICRON CMOS TECHNOLOGY

K.TORKI, S. SAULNIER
TIMA-CMP Laboratory
46, Av. Félix VIALLET, 38031 GRENOBLE, France
Kholdoun.Torki@imag.fr *Sebastien.Saulnier@imag.fr*

Abstract

The paper describes some concepts on timing modeling and the way it has been used for porting a complete digital library from silicon input data to a commercial CAD tool simulation data. This task of porting libraries into CAD tools is one of the expertise field of CMP. Academia, research and industry are taking benefit from those software developments for circuit design and training.
For CMP it is important that a large number of CAD tools is supported for a foundry technology. (The foundries usually focus on only a limited number of CAD tools, because of the support that this involves).
This paper is describing only a part of the development, and concerns the timing implementation of a sub-micron digital standard-cell library. The complete development included all the steps of the design-flow (front-to-back), including synthesis, simulation, schematic capture, floor-planning, automatic place & route, back-annotation, LVS, DRC, extraction, and full-custom technology files and environment.

1 Introduction

The figure 1 describes the complete design-flow that is suitable for standard-cell design. Each step of the design flow corresponds to the use of a CAD tool. It allows to perform an operation that help the designer to solve the problem of integrating the design until the last step which is the verified layout that has to be fabricated.
The task of porting a standard cell library into a CAD tool consists of describing the different representations of each cell into the CAD software internal language. Each of those representations will be used by the CAD individual tools (synthesis, simulation, automatic place & route, etc ...) for design. The resulting portage is called "design-kit". CMP developed in the past several design-kits, some of them are used actually by users in academia/research and industrial entities, and are considered as products. In other words, the design-kit can be considered as a configuration of the CAD software for a given technology and library. It is essential for a designer to have access to a design-kit compatible with the CAD functionality and compatible with the foundry specifications.
In this paper we describe how we faced with the problem of porting a sub-micron standard-cell library timing specifications into an industrial CAD tool that supports the “input slope model”. We started from a "Non-Linear Model Table" description of the standard-cell timing. The original specifications of the standard-cell library provided by the foundry includes a "Non-Linear Model" for each cell. We established the different equations and algorithms to calculate the different coefficient for the "input slope model" from the "Non-Linear Model Tables" and qualified twice the results by comparing the theoretical and simulated results. The comparisons give only 455 erroneous values among 54000 total parameters and coefficients. This represents about 99.157% success of our calculations. The remaining 0.843% erroneous values has been manually corrected or intentionally ignored, since they affect the timing of some particular situations were the accuracy is either not relevant or has been corrected manually. All the erroneous values has been reported and evaluated and the main reason of those errors concerned pad cells, tri-state logics, and some limit case like maximum slope and minimum load.
In the following paragraphs, we summarize the basic principle of the "input slope model" and we give an example of the non-linear table model provided by the foundry.
The full report on the timing calculation and qualification can be obtained on request.

2 Calculation method deducing propagation delay coefficients from non linear characterization

The logic and timing simulation made during the pre-layout phase and the post-layout phase verifying the effects of placement and routing on design performance, estimate the propagation delay.
The propagation delay in a circuit path is calculated as the sum of delay segments in the path.

T.J. Mouthaan and C. Salm (eds.), Microelectronics Education, 239-242.

A segment delay in the path depends on 4 components:
- intrinsic delay
- delay due to capacitive loading at output
- delay due to transition time for the inputs
- interconnect delay

The propagation delay of each cell can be calculated using a linear model as the sum of the delay components (except interconnect delay) or interpolated from a look-up table.

3 Linear model

The linear model of the propagation delay is approximated as:

$$tD = [t_{\text{int}} + t_{drive} + t_{inputSlope}] \times PVT$$

Where : - tD is the total propagation delay through a cell.
- t_{int} is the intrinsic (rising or falling) delay of the cell.
- t_{drive} is the drive delay caused by the capacitive loading at the output.
- $t_{inputSlope}$ is the input slope delay caused by the slope of the (rising or falling) input signal.
- P,V and T are scaling factor to take into account any change in process, voltage and temperature conditions.

3.1 Intrinsic delay

The intrinsic delay t_{int} is the delay caused by the cell itself when it propagates a signal from a given input pin to a given output pin. (Figure 2)

3.2 Drive delay

The drive delay results from the limited ability of an output to charge or discharge its capacitive load. To take into account the dependency of the drive delay on the capacitive load, it is calculated as follows:

$$t_{drice\_ri\sin g} = D_{ri\sin g} \times C_{load}$$

$$t_{drive\_falling} = D_{falling} \times C_{load}$$

Where : - D_{rising} is the drive delay coefficient for a rising output (ns/pF).
- $D_{falling}$ is the drive delay coefficient for a falling output (ns/pF).
- C_{load} is the total capacitive load on the output (pF).

To characterize D_{rising} and $D_{falling}$ we vary the capacitance load on the output and measure the propagation delay. D_{rising} and $D_{falling}$ represent the linear slopes estimated between the cell delay and the loading relation for rising and falling signal transitions, respectively, as shown in figure 3.

To improve the accuracy, some models take into account the variances in output load dependency using separate sets of parameters for each capacitive region. The figure 4 is an exemple with two load dependency regions.

3.3 Input slope delay

The input slope delay is the delay due to transition time for the input, it means the time it takes the input signal to change between two logical states. The figure 5 shows how the input slope can affect the output signal. The left signal in figure 5(a) has a perpendicular slope that corresponds to a step function applied at the input. The left signal in figure 5(b) has a large slope that corresponds to a slow rising signal. The outputs are shown on the right of the two figures. It takes the inverter much longer to drive its output to zero if a slow rising input is applied.

Typically, we use a linear or a two-piece linear model to approximate the effect of the input slope.

3.4 Input slope linear model

With a linear model the input slope delay is calculated as follows:

$$t_{inputSlop\_ri\sin g} = S_{ri\sin g} \times IS \qquad t_{inputSlope\_falling} = S_{falling} \times IS$$

Where : - S_{rising} is the input slope delay coefficient for a rising output.
- $S_{falling}$ is the input slope delay coefficient for a falling output.
- IS is the input slope at the input pin of the cell (ns)

To characterize S_{rising} and $S_{falling}$ we vary the input slope and measure the propagation delay. S_{rising} and $S_{falling}$ represent the linear slope estimated between the cell delay and the input slope relationship for rising and falling output signal transitions, respectively as show in figure 6 .

The input slope depends on the output slope of the driving cell so we need to calculate the output slope to be able to propagate this result along the path. It is a linear function of load.

$$OS_{rising} = OS_{0\_rising} + O_{rising} \times C_{load}$$

$$OS_{falling} = OS_{0\_falling} + O_{falling} \times C_{load}$$

Where : - OS_{0_rising} and $OS_{0\_falling}$ (ns) are the output slopes of the cell without any load. To measure OS_{0_rising} and $OS_{0\_falling}$, we apply a step function to the cell with no load attached to it.
- O_{rising} and $O_{falling}$ are the output slope coefficients for a rising or falling input (ns/pF). To characterize it, we vary the output capacitance and measure the output load and make a graph as figure 7 shows.
- C_{load} is the total capacitive load on the output (pF).

4 Non-Linear table model

Other method to estimate the propagation delay is to interpolate it from a table. To characterize the cell delay non-linearly, we can vary the load and the input slope and plot the result in a table. The output slope can be another table model based just on the load or both input slope and the load.
The data provided by the foundry to us have had this form, and we used them as input data for the "input slope model" coefficient calculation.

a) propagation delay table

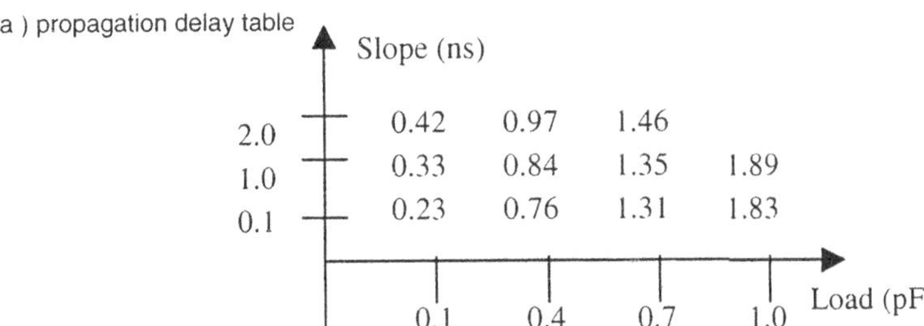

b) one dimensional output slope table

Load (pF)	0.1	0.4	0.7	1.0
Ouput slope (ns)	0.73	2.02	3.42	5.03

5 Conclusion

The timing implementation of about 300 standard-cells have been made on a CAD tool that supports "input slope model" for the logic simulation. The data used are provided to us in form of "Non-Linear Tables" by the foundry. The implementation gave automatically a 99.157% success (for less than 10% timing error). The remaining 0.843% erroneous values has been automatically identified and manually corrected. Figure 8 shows the procedure used to qualify the timing information data.
This task of porting libraries from the foundry documents to the CAD software needs an in-depth expertise both in process, libraries and CAD.
CMP is providing the result of those developments to Universities, Research Laboratories and Industry, for designing high performance circuit. CMP by those developments (design-kits) ensures the link between the CAD tools and the circuit fabrication.

Acknowledgements :
The authors would like to thank Austria Mikro Systeme for their input data and technical support.

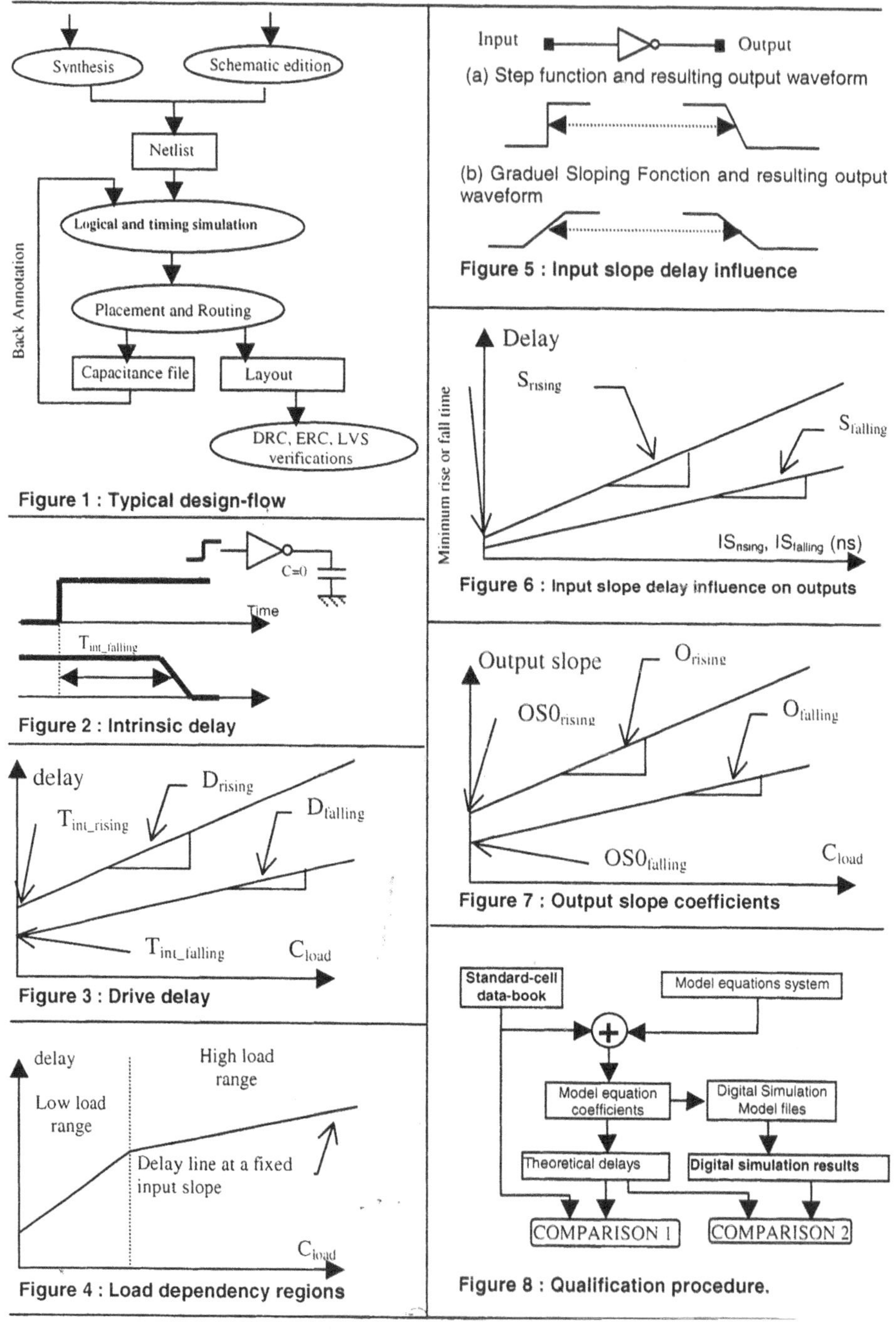

Figure 1 : Typical design-flow

Figure 2 : Intrinsic delay

Figure 3 : Drive delay

Figure 4 : Load dependency regions

Figure 5 : Input slope delay influence

Figure 6 : Input slope delay influence on outputs

Figure 7 : Output slope coefficients

Figure 8 : Qualification procedure.

A VHDL COURSE BASED ON THE EVITA™ MULTIMEDIA TUTORIAL

J. MIRKOWSKI, M. KAPUSTKA, Z.SKOWRONSKI,
A.BINISZKIEWICZ
Technical University of Zielona Gora
Dept. of Computer Engineering and Electronics
ul. Podgorna 50
65-246 Zielona Gora, Poland

Abstract

This short paper introduces the EVITA multimedia tutorial – a new learning tool for beginners in VHDL. An outline of a VHDL course based on EVITA and other tools from the Active series (offered by ALDEC, Inc.) is presented.

1. Introduction

Undoubtedly VHDL is playing an increasingly important role in hardware design nowadays. A natural consequence of this fact is the need for good, effective and accurate courses of the language aimed both at universities and individual learners. Such a comprehensive course, preferably integrating course material and software tools, could be a great help, especially for a newcomer to the field, where over 80 book titles and over 80000 web pages can indeed be overwhelming.
This paper presents an attempt to support potential users – both course organizers and individuals – with a comprehensive VHDL course based on a new multimedia tutorial EVITA. EVITA is not only a tutorial, but also a language reference guide with over 200 examples and an almost 300-page workbook. Finally, the course assumes the use of a PC-based Active-VHDL simulator and an Active-CAD synthesizer, all from the same vendor.
A multimedia tutorial for VHDL language is nothing new – there are at least four on the market now [1, 2, 3, 4]. They are well written, with useful (however compact) Reference Guides and a workbooks of limited usage. In our opinion they lacks interactive experiments as the user can only switch from page to page, browsing supported information.
All these observations led to the conclusion that there is a need for a new multimedia tutorial, where the disadvantages of mentioned products are overcome: interesting graphics, animations, on-line experiments (changing values, simulations), language reference guide and a printed workbook that recalls the information from the screen and supports the user with additional examples and space for personal notes. These were the preliminary assumptions that laid down the fundamentals of EVITA™ [5].

T.J. Mouthaan and C. Salm (eds.), Microelectronics Education, 243-246.

The tutorial itself is presented in Section 2. The next Section briefly introduces other software tools that can be used for the practical part of a VHDL course based on EVITA. An outline of such a course is presented in Section 4. Section 5 concludes the paper and brings comments on EVITA's limitations as well as plans for improvement.

2. The Tutorial

Keeping in mind the additional mechanisms and tools implemented in Evita, we can state this product represents a new point of view on education process. First of all Evita is the one and only tutorial that allows user to interact with it on such a scale. The user can have influence on the parameters of VHDL programs, timing parameters with instant simulation, influence on source code or can input source code and check its correctness. Such interaction possibilities are implemented in most pages and this is exceptional in comparison to similar products on the educational market.

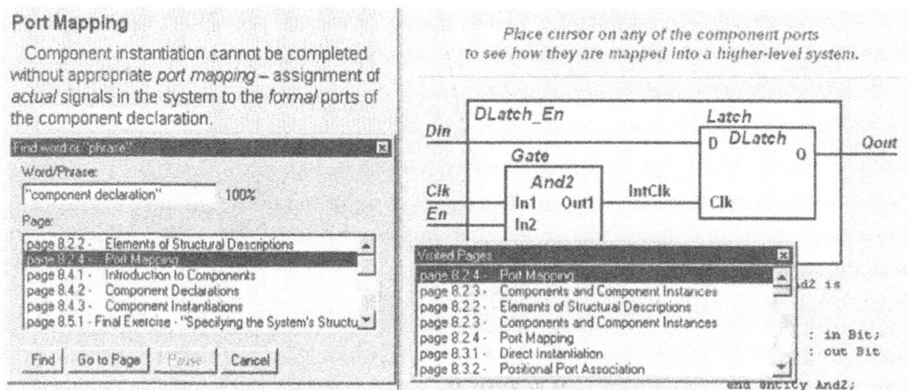

Figure 1. Tutorial in the middle of work.

Material presented in the workbook is divided like books (chapters, sections and pages) + additional pages as in the chapter summary where the abstract of chapter is presented. The even page in each section is divided into two parts:

- **KEY ISSUES** - contains the same text presented in the module;
- **EXAMPLES** – apart from material from the tutorial (text + picture) the user has an additional example which explains the topic in a more informal way.

The odd page in each section is divided into three parts:

- **ILLUSTRATION** - the same illustration that is used in the module;
- **REFERENCE GUIDE KEYWORDS** - contains a list of keywords used on the tutorial page which are available in the VHDL Reference Guide;
- **USER NOTES** - allows to put notes taken from work with the module.

The "User notes" part is very important because with the completed "User Notes", the workbook becomes full a VHDL manual for beginners. Please keep in mind that the workbook is ready to use immediately. The user does not need a computer, screen or desk but his only knees, so access to the data in the workbook is simply rapid.

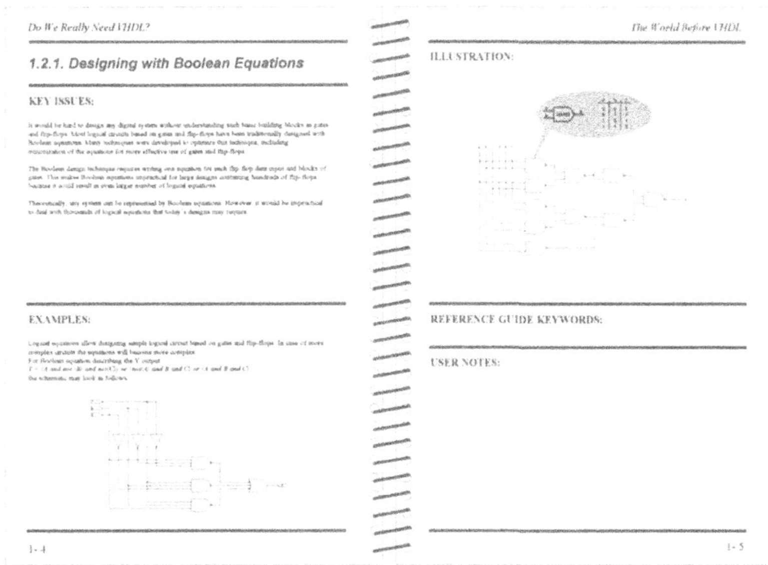

Figure 2. One workbook section equal to one module page.

The next tool implemented in the EVITA VHDL Tutorial includes the same Reference Guide help (RG) as in Active-VHDL. The RG consists of 82 VHDL keywords and over 200 examples. A page in RG is divided into 5 parts: Formal Definition, Description, Important notes, Simplified Syntax, Examples.

3. Other Tools Used

EVITA was created to supplement work with Active-VHDL VHDL simulator. This software is a VHDL centric design environment that provides a broad range of resources, quality, speed, on-line help, training materials and documentation. Active-VHDL provides tools for efficient design management, design entry and simulation at all levels of design development. You can test your designs behaviorally and perform timing simulations after synthesis and place & route. Active-VHDL provides interfaces to external programs such as synthesis software or place & route tools. You can, therefore, synthesize designs and perform place and route within the same environment by executing macro command files which allows for closer cooperation between Active-VHDL and other synthesis tools like Altera-MAX+Plus II, Cypress-Warp2, Xilinx-FPGA Express and others. Because of Activ-VHDL flexibility designers have a very good, easy to use environment for designing and learning more about the use, context, examples and usage of the VHDL language.

4. Application to a VHDL Course

The learning process can be divided into two parts: the first step is to gather information e. g. through the lectures. At the same time the student can refer to the page in the module on which he can find additional examples, usage and notes concerning material presented by the lecturer. This more theoretical stage is followed by the second step, which involves the use of nearly acquired knowledge by the student. In this part the student can fully absorb the information through exercises, tests and a generally hands-on approach. Finally, the student can assess his own progress through tests, which simultaneously deepen his knowledge. In the case of no answer being specified by the student, the module advises him where to go to find appropriate answer.

5. Conclusions and Further Developments

Active-VHDL and EVITA appeared on the market only recently (Q1 1998) and as a result any experiences in using it for educational purposes are quite limited. Nevertheless, due to close cooperation between ALDEC, Inc. and the Technical University of Zielona Gora we have had an opportunity to use preliminary versions of the tools before their official release. The results (student's opinions on the course) are quite promising for the future. In its present form EVITA can serve as an excellent introduction to VHDL and as such (especially due to extensive, well-illustrated user interactions and top-quality workbook) can certainly rank among the very best multimedia tutorials available. Nevertheless, there are two areas where it can be improved:

- as introductory material, it lacks discussion on more advanced topics like subprograms or declarations of packages. Now they are present in the reference guide only.
- Interactive cooperation with the Active-VHDL simulator would enhance learning capabilities very much and is highly desired by the users.

Both features will appear in the extended and revised version of EVITA, which will be available, later this year.

6. Acknowledgements

Part of this work has been done when the first author was with the Department of Computing Science, University of Newcastle upon Tyne, UK on a Visiting Research Fellowship sponsored by NATO/Royal Society.

EVITA, Active-CAD and Active-VHDL are registered trademarks of ALDEC, Inc.

7. References

[1] http://www.esperan.com
[2] http://www.doulos.com
[3] http://www.whdl.com
[4] http://edec.brookes.ac.uk
[5] http://www.aldec.com

INTRODUCTION TO FPGA BY TRAINING EXAMPLES - APPLICATION IN A PCM SYSTEM

F.NOUVEL, C.MOY, O. PAVIOT
Laboratoire Composants et Systèmes pour Télécommunications
Centre Commun Micro-électronique de l'Ouest
UPRESA 6075, INSA 35043 Rennes Cédex FRANCE
Tel : 99 28 65 07 Fax 99 38 62 48
E-mail : Fabienne.nouvel@insa-rennes.fr

1 Abstract

This paper describes how FPGA technology is introduced to the 3ESC and 4ESC students. ESC means ' Electronique et Systèmes de Communications '. This option of the INSA school has been created for two years and prepares senior engineer students in 3 years. During the courses and practice, students have to realise a correlator system, using different entries. In 4ESC, they work in small groups (6 students) and perform a complete electronic system, based on FPGA. The two systems are presented in this paper and the advantage of FPGA, using ALTERA.

2. Contents of the course

In 3 ESC, the CPLD/FPGA technology is presented to the students. Different aspects are presented :
- first , the components with their advantages and disadvantages
- the PLD/PLA architectures (including RAM/ROM)
- the CPLD architecture, with the 7000 family of ALTERA
- the FPGA architecture with the Flex 8000 family of ALTERA

This part is presented during courses, of about 14 hours. Other families are also presented, in order to give the students a wide overview of the FPGA/CPLD technology. Then , the programmation and methodology of implementation of the components is introduced :
- the schematic entry
- the hdl(hardware design language) entry
- the simulation and tests of the realised system

This part represents six labs of 4 hours (24 hours) in 3ESC. This part is very important because students learn how to use a tool (ALTERA in this case) ; they need to have a top-down approach (the hierarchical identification of different sub-systems) ; they have to test the system in different components (clpd, fpga) in order to see the advantages of each. In 4ESC, when they work alone, they have to take into account all what they have learn. They have to choose the components (the chipper), the approach (schematic, hdl , ...).

3. The choice of ALTERA

Many tools are present on the market. In the case of education, we need to have a tool :
- simple of use and easy to use, fast intuitive in order to maximise learning
- a complete integrated design flow
- a on help line (necessary when the students work alone)
- the cycle of development must be short, in order to give rapidly a response to the students. So design entry, processing, and verification together must take just a few hours, with several complete iterations in one lab. Many features are available (as in industry)

T.J. Mouthaan and C. Salm (eds.), Microelectronics Education, 247-250.

In our case, we have take ALTERA tool, with Maxplus2, on Windows95. ALTERA and the CNFM have signed convention in order to install the software and hardware in different universities. This facility permits to use recent tools, widely used in industry, with a low cost.

4. Description of the 3ESC project (labs)

During the labs, students have to realise a programmable correlator system.

4.1 THEORY

The correlation of two random codes is used to compare these one, in order to measure, for example the multipaths propagation in radiocommunication, the radio channel characterisation, in CDMA system (Code Division Multiple Access). This function is given by :

$$R_{xy}(k\cdot T)=\sum_{n=-\infty}^{n=+\infty}x(n\cdot T)\cdot y((k+n)\cdot T)$$

T represents the period of the function. This equation can be written

$$R_{xy}(k)=\sum_{n=-\infty}^{n=+\infty}x(n)\cdot y(k+n)$$

If x and y have a period of Lc, this function R_{xy} has also a period of Lc. So, Lc samples are necessary to calculate the correlation.

$$R_{xy}(k)=\sum_{n=0}^{L_c-1}x(n)\cdot y(k+n)$$

By observing the equation, different sub-systems appear :

- the generation of the pseudo-random codes . One is fixed, the second one must be translated at each end of period Lc (effect of k).
- the correlation function, which is an accumulation
- the memorisation of the result

4.1 REALISATION

The system is presented on figure 1. Each sub-system is presented as a symbol, with a schematic or hdl representation.

4.1.1 *the pseudo-random generator*

This generator is presented on figure 2. A pseudo-random code is characterised by : the length of the code Lc, the choice of the configuration. The number of flip-flop or registers used is given by N, where :

$$Lc = 2^N$$

The configuration represents the flip-flop used to generate the function. The user can select the length and the configuration of the code. This generator is duplicated, as a generator for the transmission and a generator for the reception

4.1.2 *the state machine*

The state machine is used in order to synchronise the initialisation phases of the system. The user has to give the different parameters : length of the codes, configuration of the two codes, beginning of the correlation. The state machine is described in AHDL (Altera HDL) language, which is easy to use, and very close to a sequential description.

4.1.3 *the correlator*

The correlator is an accumulator, which is incremented when the 2 bits of the 2 generators are the same. When the 2 codes are the same during all the sequence, the value of the correlation is maximum and equal to the length of the codes. When they are different, the value is -1 (in the case of code with maximal length).

4.1.4 *the simulation*

At each state, students have to test the system. They first simulate the sub-systems, functionally (they don't take into account the component). Then, they compile and simulate it in a component (the timing simulation). They can observe the differences between the different families, cpld and fpga. By using the floorplan editor, the designer views all assigned and unassigned logic in the device (pins, cells , ...). They can observe the influence of manual assignments on the performances. This part is important to determine the best solution of the system.

5. Description of the 4ESC project

During the 4ESC year, students have to realise a complete integrated system, a PMC system (Pulse Modulation Code). This technique is used for the numerical phone. This system includes the transmitter and receiver.

5.1 DESCRIPTION

In the transmitter, 4 channels are used, three for the digital compressed voice and one for the synchronisation(in a real system, we have 32 channels). The channel are multiplexed in time, with a 8khz sample frequency. The transmitted data can be replaced by a pseudo-random code, used to test the transmission. The information is coded, according the HDB3 format (without DC value, limited bandwidth, multiple transitions for synchronisation). In the receiver, the 4 channels are separated, decoded. An important sub-system is the synchronisation module, necessary to re-find the good timing and duty cycle.

5.2 REALISATION

5.1.1 *the transmitter*

It is presented in figure 3. Different approaches have been used. For the upper level, a schematic entry is used. For the compression and decompression law, all have been written in AHDL (it is represented has a conversion table). The multiplex function is designed in schematic (it generates the order for an analogical multiplex circuit). The HDB3 coding is performed in schematic (AHDL can be used).

5.1.2 *the receiver*

It is presented in figure 4. All the symmetrical parts of the transmitter can be identified, with the synchronisation part more. The two parts have been tested separately, functionally and timing (on a flex 8000 component). Now, all the system has to be tested.

6. Conclusion

Different realisation on FPGA/CPLD have been presented. The ALTERA MAXPLUS2 is used to introduce this technology to students. This paper shows that students can perform interesting projects, and are rapidly autonomous.

We would like to thank ALTERA, Europractice and the CNFM for their contribution for the tools.

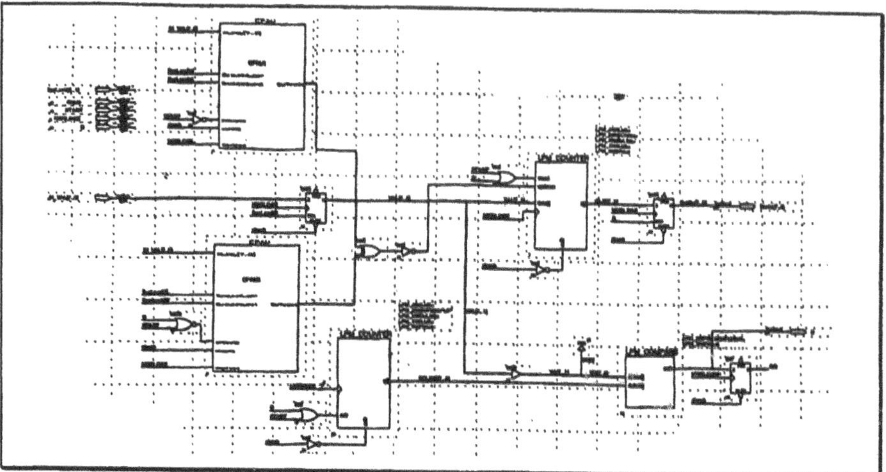

Figure 1 : correlator system with the 2 pseudo-random generators

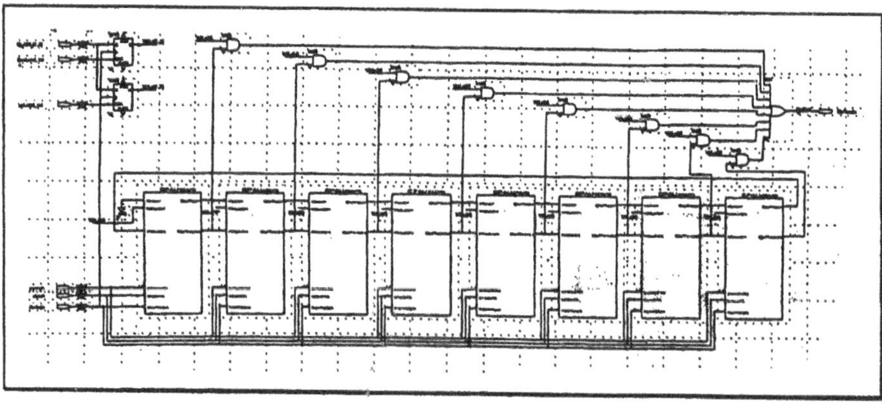

Figure 2 : pseudo-random generator, programmable

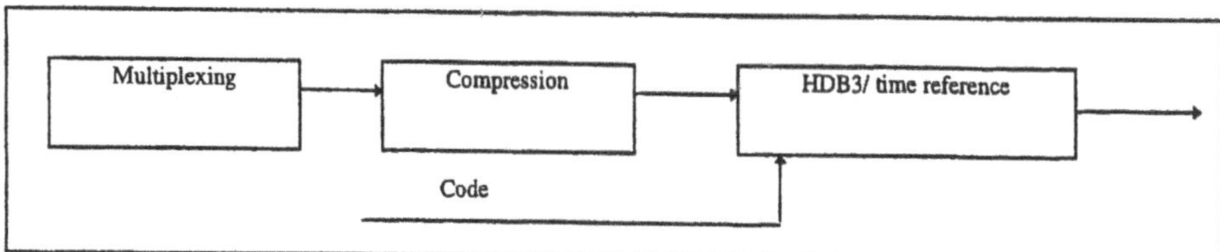

Figure 3 : the PCM transmitter

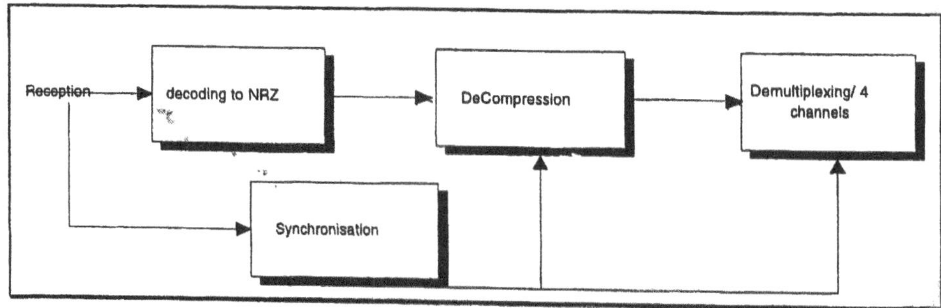

Figure 4 : the PCM receiver

Session E

MULTI MEDIA

WRITE AND USE HYPERMEDIAS FOR MICROELECTRONIC TEACHING, HOW TO TAKE UP THIS NEW CHALLENGE?

Dr. A. Galisson * & Dr. E. Zysman **

* ENST- departement electronique- 46, rue Barrault, 75634 Paris cedex 13
** EPFL - LEG, CH - 1015 Lausanne, Switzerland

Abstract

The evolution of the microelectronic domain marks a turning point with a dramatic growth of human resources. A mass of qualified personnel must be trained in diversified areas, more and more their training may long their whole professional life. Hypermedia techniques will widely contribute in the close future to teaching methods. This new challenge could turn very inefficient and costly if some objectives are not satisfied. Thus, the identification of the public concerned becomes fundamental to garantee a suitable training, a complete transformation of the lecturers and the courses "creators" will be necessary to engage them on the Hypermedia way, the design of these products will involve a wide range of constraints including fundamental and human sciences.

1. Introduction

1.1 Microelectronic trends

The microelectronic domain has entered in a new era. Technology will continue to progress, but its maturity and performances suggest now inumerable and ambitious applications. Ideas in circuit design and application must be prioritized both in digital or analog domains. These trends will require new specialists in electronics, physics, computing science.
As a key sector of activities in the worldwide economy, an efficient teaching with targets correctly identified would turn necessary, in order to produce an increasing number of human ressources who will:

- demonstrate more autonomy,
- work with a large tool palette (logical simulator, electrical simulator, design tool, high level description tool, analysis and test tools, etc..)
- adapt continuously and fastly their knowledge and skillness owing to the permanent evolution of the domain.

1.2 New challenge for industries and countries

Microelectronics as computing did in the past, will flood industries, in most sectors of activities. Creating specific circuits will turn common:

- Microelectronics is very close to computing. A digital integrated circuit can be assimilated to an algorithm integrated on silicium.
- Microelectronics in term of design becomes possible today. Generate a circuit becomes easy if we consider that adapted and cheap tools already exist.

The technology challenge will turn harder in the future with North America and South-East Asia. Even though, we cannot stay inactive and dependent in terms of know-how. The implications of microelectronics are too big to give in. We know how education is a sensitive domain to affirm the cultural identity, and then the independance of a

T.J. Mouthaan and C. Salm (eds.), Microelectronics Education, 253-260.

country. It is then essential that Europe would contribute to high level training of its proper students, assuming then its role of decision-maker. Whatever is the orientation of an industry (evolution of the technology, new market products, creation of new sectors of activities), competences must be created or adapted.

2.- Training increasing human resources

2.1 Who are concerned?

The future will probably belong to those who will master the communication, through the performances of the means and the quality of the informations. It is then a challenge at the same time technical and human. In term of education it will be verified too.
To satisfy these constraints the public must be defined. As a young domain, the concerned public is not fixed. New teachers must be operational, graduated technicians or engineers must be formed, industries have to adapt themselves, evoluate or even reoriented implying a larger vocational training, post-graduation courses must be proposed in order to examine permanently the state of the art and prepare new specialists in a domain where any delay turns out prejudicial, high school students must be motivated

2.2 The promotion of this area

As we said before, the needs in various disciplines will increase. It means that the promotion of this area must be effective as any marketing approach to seduce new vocations. Training begins with motivation and good perspectives for the future in terms of employment and interest. Diffusion through Internet would be beneficial including links to some sites of industries involved in microelectronics. The message must convince, that the market already exists, the expansion of the domain is real, for a durable time. It must show that many competences are required, satisfying then many tastes. The economic and the intrinsic interest of the domain in term of research, development, production or in ideas and new market must be shown.

3.- Teaching microelectronics

3.1 Is it possible to propose a method?

What method(s) can fit these constraints? Today we have not a sufficient feedback to formalize the best way. We will certainly oriented pedagogy towards a mixture between traditional methods and hypermedia supports. Some products based on strong interactivity, animations, qualitative simulations already exist. Traditionnal methods will persist but turn more social because of the interaction between persons must be garantied. Thus, in the future a strong development will satisfy any didactic activity like lessons, individual revisions, indepth analysis, laboratories, exercises, projects.
Taking up this challenge could appear, from outside, to a soft evolution considering that Microelectronics is already linked to computing science through a large involvement of CAD tools used for quantitative aspects and covering the whole domain.

3.2 A particular state of mind

The future engineer will have to deal with little fatal errors, top-down approach to manage as long as possible a high degree of abstraction, local solutions which will optimize the circuit, compromises with speed, surfaces, consuming, simplicity of the concept, time realization,etc... Obviously, a global training in microelectronics means the assimilation of a mass of information and technical competences. But the acquisition of a state of mind is essential too. This state of mind will contribute to rigour, methodology, intuition, etc..., proper to an efficient engineer. Expertise with computer will be involved in design, particularly in the decision aid. Thus, the critical mind must be developped too, because a global vision would remain better than any computer expertise for a long while. A state of mind is not formalizable, this represents the main difficulty of a traditionnal method which is generally oriented to technical and quantitative approaches. We need then complementary methods with the traditionnal ones.

3.3 Training needs close links with industries and research

What are the constraints of an high-tec training? High-Tec evoluates fastly. It is then necessary to identify what is unchanging (at least during some years), but anticipation must be planned too during the training. These responses obviously come from the industries or the research laboratories. We cannot get away an evaluation of the future training with the participation of these groups.
The form of the training will be influenced by the relation unchanging - anticipation. At the date of issue of the product, showing explicitely this relation would be the best way to appreciate the celerity of the different evolutions, the weight of the reality, the hopes and the speculations in this domain.

4.- Methods based on hypermedia

4.1 Relation between technology and training

The multimedia of tomorrow with all the expected performances and creativity will not exist without microelectronic. Treatment of audio and video datas, real time simulations or animations with interactive handling. Rare are the areas where we can observe that a knowledge support contains the explanations of the proper support. It is the case, for example, of the paper and video. This singularity is still stronger between microelectronics and multimedia.
As a matter of fact, we show through multimedia that the training itself is linked to microelectronics. Video, audio, and images need special format in order to compress and decompress them efficiently. Dedicated circuits are used. Celerity of image synthesis (with models, textures and animations) and real time simulations need powerful circuits and a huge memory. This relation between technology and training will contribute to facilitate their union.

4.2 A complementary approach with traditionnal methods

It is not interesting to adapt traditionnal methods for a multimedia aspects, because it will not show anything new and will be costly without providing a significative efficiency. There is no added value if we copy any activity that a teacher can do. A teacher can talk, present images and videos, play sounds,...
The multimedia activities must introduce a form of complementarity in terms of interest, understanding, intuition, state of mind.
Aspects linked to computing features are interesting. In this case, medias are not fundamental, only the performances of the product are significative. It is true for:

- Context-sensitive aid or link between datas
- Power index
- Instant evaluation of the knowledge status

Pedagogy is fundamentaly involved to identify where we can introduce concepts that in real time are impossible to introduce in lectures and that give something more than classical methods

- Qualitative analysis through qualitative simulation
- Pedagogical virtual labs
- None observable phenomenas or difficult to explain in oral.
- Analogy with other sciences is very important in pedagogy. We would have to exploit it with dynamism (video, animations, sounds,...).

Aspects linked to psychology are very important. It concerns principally the significatif involvement of Arts, proper to the hypermedia approach. Even the teacher presents various gifts, he will never assume at the same time Mozart, Vinci, Tolstoï, Helmut Newton, John Ford.

4.3 Multimedia: a new vision?

Multimedia permit to involve all senses adapted to appreciate the training. It must facilitate, encourage, be pleasent to manipulate. The term of pleasure is indissociable of the goal of the training. The pleasure must obviously be assumed by the student but also by the teacher.

The main difficulty will not lay in the training of students. They will adapt themselves fastly. They will have already Internet and multimedia experiences in other domains or previous training. The problem is much higher for vocational training and above all for the teachers which will need adaptation. We will in point 7 to answer to this question.

4.4 Pedagogical scenario is fundamental

What justify the use of sound, image, video,... The scenario!! The scenario justifies at an appropriate moment the use of such a ressource. A dynamic scenario will not construct the knowledge of the student. Construction means a method used in the classical method. Student must feel total freedom. The scenario will not put the student in a classic situation because with identical references he will not find the differences. Scenario is a key parameter because can get rid of systematic and austere structure (phenomenology, explanations, models, applications, ameliorations) but respecting the scientific containts.

Pedagogical scenarisation is important because:

- It is the reflect of creativity in terms of pedagogy
- It conciliates creativity and feasibility, and then fixes the cost to invest.
- It introduces easily study cases, merging observation, analysis, understanding with transparency related to methodological constraints.

Assuming that hypermedia will not substitute teachers, pedagogical scenario must complete the vision of the classroom or even propose an opposite method. In the class the orchestra conductor is the teacher. His approach is generally formal, showing, explaining, modelizing and using phenomenas. He has a goal to reach, a program to respect, a defined timing to do it, in resume a mission and the student is oftenly passive. At home the student is alone with his problems, his gaps and his own motivation.

It would be important to propose an environment which doesn't take into account the possible difficulties but maintain the **student motivation**. "Heretic" vision can be propose, for example educationnal games where the students plays the role of a searcher who discovers as a true scientist. The multimedia would be a subconscient guide of the student.

4.5 Qualitative approach in hypermedia

In microelectronics, designers use to work with computers environment just with CAD tools. Hypermedia in microelectronics will not represent a revolution for users. However, the users work generally with quantitative elements in mind: Exact surface, exact time, logical simulation. Electronics doesn't support "more or less". But still, to master any domain like the complexity in microelectronics, a top-down approach is necessary. It introduces at each level abstraction levels which are refined as work progesses. The designer leans on a rigourous approach. This mechanical side doesn't imply that he has obviously understand all the refinement of the technology and the design. Multimedia through visual aspects, is a good way to present magnitude comparison and qualitative views. These elements are necessary to develop his intuition and the criticism mind, necessary to lead him to anticipate problems, optimize them....

5.- The constraints of this approach

5.1 Constraints implied by the microelectronic domain

It is not healthy to propose to do multimedia just for multimedia. Taking into account the cost of such an application, we must be certain to make cost-effective our business knowing that we have a lot of constraints:

- Microelectronics techniques are fastly obsolete,
- Design and technology are closer today
- We don't have to manage only technical constraints but also a state of mind ,
- English is the main (almost alone) language used in computing literature and software, while teaching is generally made in the local language.

5.2 Hypermedia is expensive

Multimedia is expensive. According to the nature of the project and its ambition, a multimedia product can present a charge of several million dollars. The cost is justified if we admit that the realization of such a product supposes that:

- the course is completely covered.
- the product has a high quality, without error or prohibitive waste of time.
- the scientific message is accessible, easy to understand and pleasent to follow.
- subjective aspects are necessary to improve the presentation
- motivating and pedagogically efficient for the students.

Since the proper discipline to train and the different techniques involved in hypermedia are mastered, it is possible to virtually represent anything. The limits of hypermedia are those of our imagination and mainly our wallet. All didactic activities can then be satisfied like lectures (showing phenomenology, explaining it, modeling, application, optimisation and refinement) exercises, laboratories and projects, self evaluation.

Multimedia is costly because everything is possible. We can virtually represent anything with sounds and images. Why not imagine an interactive and animated show like a "Silicon Park", where electron will sing, after marrying a proton, and during the same time observes the substrate contorting by pain during thermical diffusion, etc... It is obvious that pleasure and a play activity are indissociable parameters of the training and then of the future educationnal hypermedia. Unfortunately, it could turn an inconsient (or conscient) pretext to do beautiful and creative applications that would have a weak didactic impact but a prohibitive cost.

The features above indicate that it is not possible to work with limited groups, even more the teacher has turned a, complete scenarist knowing well what we can do in multimedia in order to conciliate creativity and feasibility. Several professions must be

implicated. In technical discipline like microelectronics, the human component should miss inside hypermedia support. Without any external "guide", the product could turn difficult to use. Human sciences must be involved during the design phase of the product in order to turn it more "intelligible". Besides the pedagogue and the computer scientist implementing the product, we must cover aspects like ergonomy, semiology, psychology, arts for all the non-scientific criterias which will contribute to the quality of the product. Juridical aspects have to be taken into account too for all the problems linked to the rights.
Observing the size of such a team, it would be preferable to engage immediately a collaboration between universities and industries. As we consider that microelectronics has turned supranational, training will do the same. There is no other alternative than cooperation.

6.- some compromises to accept

6.1 Waste time for the correct questions

A good teaching assumes that essential questions must be resolved. An exhaustive list of questions is impossible to lay down. However we can examine the following:
- What material can be proposed knowing that what is teached today is already obsolete, technology and design have now closed interaction and do not constitute independant areas, digital and analog circuits show similar constraints when performances are reached,...
- What know-how and "spirit" must be transmited. A critical mind is necessary taking into account that anything in microelectronics has a short-lived period. Rigour based on an efficient methodology is better than an amount of informations. Developing autonomy is necessary to absorbe each evolution. In resume quality instead of quantity.
- Can we find universal material in order to enhance our chance to create a product which will not turn fastly obsolete. (like basis in electronics)
- Does our pedagogy bring up something new in term of pedagogy. Doing the same is a loss of money.
- We must identify to which group the training is dedicated. We will not train technologue, Designer or specialist in CAD tool development with the same approach neither the same state of mind. Inside a group, what are the professionnal maturity, the knowledge already acquired (in the domain and in multimedia approaches) the environment where the group is, etc...
- What are the priorities of a training: transmit a knowledge, transmit the capability of adaptation, transmit the capability of autonomy, transmit the motivation, etc.... Various are based on the psychology of the future engineer.
- How can we garanty the update of the knowledge. Mixed solutions between CD-ROM and Internet look reasonnable. CD-ROM will include the largest part of the materia, while Internet will update regularly the new datas.
- Do we need to follow a method or leave this rigor in classroom and promote an free approach. A guide who wants to be understood must be structured, methodic.
- Is the product dedicated to learn or consolidate the knowledge.
- How diffuse the information. We know that the quality of the support is fundamental. However, a CD-ROM is limited for a wide diffusion and to cover all a domain. Internet is too slow.
- Microelectronics is settled between exact science and applied science. The two approaches are different. What approach will be priviledged
- Other questions are related to Economical interest (what is better to do?), Philosophy (What would I like to research?), Practice (How can I manage with the specification I have?),...

6.2 No future without collaboration

A large group must be constituted to turn possible the fast realization of a product, covering the largest domain as possible. Intermediary solutions are always frustrating the users.

This solution is the only way to

- minimize the costs
- specialize the efforts
- establish a competence network (aspect technical and human). Its use will allow to confront various visions and respond to the fundamental questions of pedagogy. In the future, it will constitute the node of projects covering different domains of microelectronics.
- confront complementary visions. There is no universal pedagogy considering that it is linked to socio-cultural aspects

7.- Future teachers

7.1 The teacher difficulties

Perfection is the holy Graal quest of any pedagogue. It means a dual sentiment of stress, due to the huge responsability towards his listeners, and self-pleasure when his qualities are recognized. His quest means continuous efforts and evolution. The technology progress is significative while our pedagogical culture, derived from the cartesian discourse, evoluates with a relative slowness.

Teachers will be confronted in the future to three alternatives:

- They feel a visceral reluctance to any technology support for their discipline, considering that pedagogy remains an excusive human activity. It is not convenient to blame teachers who consider pedagogy as the reflection of their personnality. Anyway their role will remain essential.
- The teacher will prefere an individual approach, using author systems and some cheap tools that are available today. Multimedia product developped with this state of spirit will stay artisanal. In microelectronics where evolution is dramatically faster than a single man production, this choice is doomed to failure if not oriented for basical components or concepts studies.
- The third alternative will turn hypermedia as a collective business. It means an organisation managing human ressources, documentation centre, training of the proper teachers, accounting, etc..., in resume a structure of an industry dedicated to multimedia development. As microelectronics, this activity will turn supranational, only manner to take up the challenge of the new millenium.

7.2 A new vision of the teacher

A teacher could be genial or bad, loved or hated. As for an actor, the sentiments shown for him are generally the expression of a moment, circumstances, situations that are not objectively analyzed. Nevertheless, he has the ability to evoluate and to adapt himself. The multimedia product has not the same perspectives, not so fastly. It means that the actor activity of the teacher will remain.

Moreover, due essentially to our social dependency and our cultural patrimony, teachers will have intense activities and for a long period. The quality of their involvement, as today, will have a fundamental impact on the quality of the future scientists and then of our economy. They will not only assume a role of lecturers but also of creators, taking into account the the feasibility of new hypermedia techniques and the new supports (CD-ROM and Internet) taking the limits of our imagination further still.

8.- Conclusion

In fact, it is possible we assist to a profound change of the creators and lecturers mind. This enterprise is very huge and will suppose professionnal structures dedicated to hypermedia scripting. They will not only include pedagogues and computing engineers, but also scenarists, artists, specialists in semiology, ergonomics, juridical aspects, etc.. In resume a team which size is comparable to a movie team. We can't get away from collaboration between different universities and industries. And if we consider that Microelectronics activities has turned today supranational, as a corollary, teaching must turned supranational too.

At this moment of hypermedia development, the technical know-how exists and we can speculate on the great perspectives it represents. Today, we are probably able to create everything, but unfortunately "anything". Since the whole discipline must be created, the "anything" could turn a desaster. As we can see, we are already at the step of resolving a lot of questions linked to the efficiency and the viability of the product.

THE EU TELEMATICS APPLICATION PROGRAMME "MODEM" RESEARCH PROJECT; A EUROPEAN MICROELECTRONICS TELEMATICS BASED EDUCATIONAL INITIATIVE

G. M. Crean, N. Cordero, S. Lidholm and M. O'Sullivan
National Microelectronics Research Centre
University College Cork,Prospect Row, Cork, Ireland

1 Abstract

The objective of this paper is to present an overview of the current status of the European Union Telematics Application Programme educational initiative in microelectronics; the Multimedia Optimisation and Demonstration for Education in Microelectronics (MODEM) project. The first phase of the project which commenced in January 1996 is scheduled to be completed in April 1998. Key project achievements to-date are detailed. Formative and summative evaluation activities are discussed.

2 Introduction

High technology industries are critically dependent on both a high level R&D spend and the availability of an educated workforce for the realisation of the new products necessary for success in world markets. The short product life cycles which characterise the electronics industry make these factors particularly important for this sector. In European as in recent US studies [1], human capital is now recognised as the limiting factor to growth in the microelectronics sector. The challenge for modern economies is to provide the skilled people to sustain growth based on expansion in the high technology sector.

The objective of the MODEM project is to respond to the above challenge and develop a telematics based European wide infrastructure and organisation to support tertiary level education and specialist training in microelectronics. An overview of specific project objectives and project partners has been published but in general terms the consortium comprises universities specialising in microelectronics education and pedagogic evaluation and companies specialising in multimedia production and simulation development [2]. In addition, the project consortium maintains a Web site with general information, MODEM newsletters, deliverables, etc. (See http://nmrc.ucc.ie/modem).

Key accomplishments to-date include the development and validation of tools for the co-operative design and production of multimedia course material, the development of a computer simulation learner environment and the development of a MODEM virtual campus.

3 The MODEM environment

To-date, the MODEM project consortium has developed facilities for multimedia-based training and education in three specific areas: Production of training material (Co-

T.J. Mouthaan and C. Salm (eds.), Microelectronics Education, 261-264.

authoring); Learning environment (Simulation-assisted Learning) and Delivery (MODEM Virtual Campus).

The co-authoring and pick-and-mix facilities developed within MODEM were completed in early 1997. The resultant environment, based on LOTUS NOTES, allows both asynchronous and synchronous co-authoring. In the asynchronous mode, the authors generate training "units" which are incorporated into a shared database. Other authors working on the same project can then access/edit the material stored in this database. In the synchronous mode, subject matter experts may also use Microsoft NetMeeting, allowing application sharing, chat-box, a shared whiteboard and also audio communication. These facilities allow co-authoring in real time. Further details are available in the public project reports which can be downloaded from the MODEM web site.

Two types of simulation-assisted learning tools have been developed inside MODEM; an Internet-based learning environment built around a commercial microelectronics modelling package and a stand alone protocol simulator.

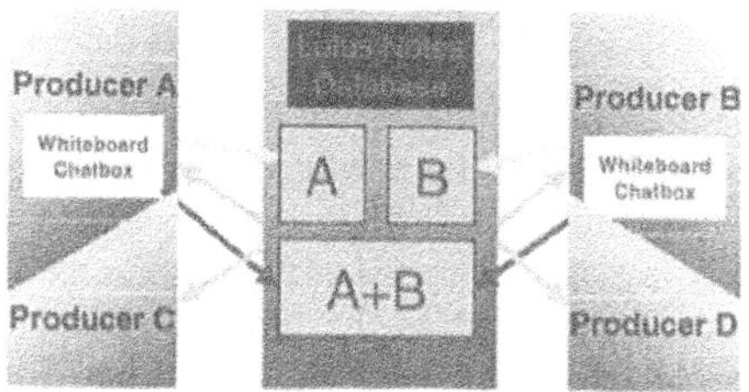

Figure 1. Co-authoring environment: interactions between course producers and the database

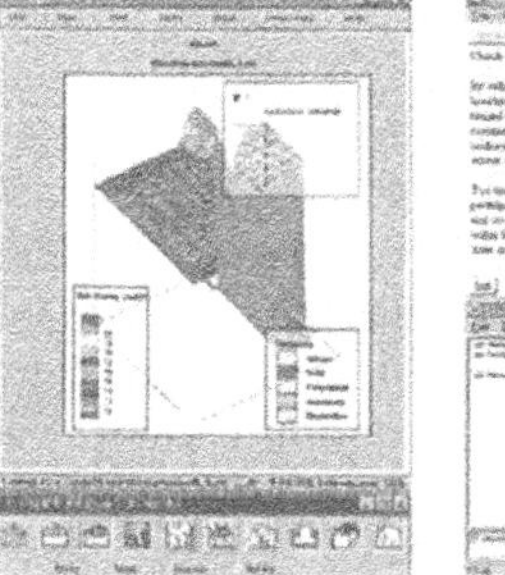

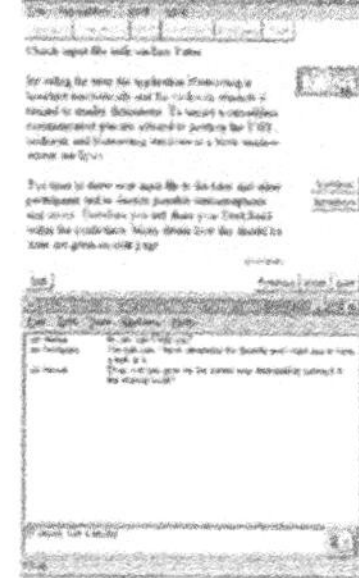

Figure 2. Simulation-assisted learning using VWF

The Internet-based learning environment [3] is built around the Virtual Wafer Fab (VWF) commercial semiconductor modelling software from Silvaco. The user first learns the design of semiconductor devices using multimedia material and then uses the simulation as a tool to complete the learning process. The integrated environment also includes co-operative learning facilities (using Microsoft NetMeeting) where several students can design an electronic device together and then run the simulation. If there is any problem, the learner/s can then contact a tutor using either on-line or off-line (e-mail) communication channels. Four tele-learning scenarios are available: Stand Alone, Tele-assistance, Tutor Supported Tele-workshop and Peer Learning Tele-workshop.

The protocol simulators are based around multimedia material demonstrators which simulate the operation of a particular piece of microelectronic materials analysis equipment (SIMS, RBS, etc.). A detailed description of the design principle has been presented [4]. These simulators can be used as part of a full multimedia training course or as an evaluation tool: a user being trained to use a piece of equipment can use the simulator to check his/her knowledge before running the real machine.

While the electronics sector has been used to demonstrate MODEM technology because of its generic nature, the simulation environment could provide a commercial advantage to European companies in any industrial sector.

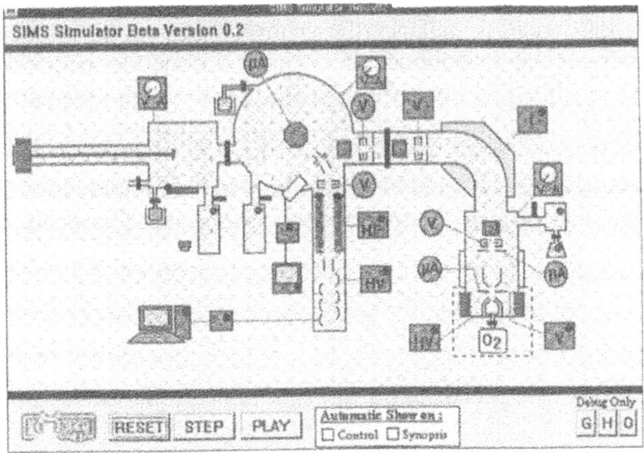

Figure 3. SIMS Simulator

Finally, the MODEM Virtual Campus [5] is the service for distribution and management of learning based on the Internet. The campus is structured like a real one, with different rooms, workshops, teachers, students, courses, etc. Users (teachers and learners) can access the course material using any standard Web browser. Every user has different access privileges (read, modify, delete), depending on the user type and registration details. There is also access to an activity server which supports services such as notice boards, help lines, workshops, etc. The Virtual Campus is installed on top of a HyperWave document server in a Unix or Windows NT system

4 Evaluation

During the course of the MODEM project, both formative (directed at improving the usability and effectiveness of the MODEM tools) and summative (used for re-active, decision-oriented) evaluation activities have been performed at several academic test sites through Europe, making use of interviews, questionnaires, checklists and observations.

The formative evaluation activities have included production and training sessions along with design walkthroughs with evaluators, producers and students. Summative evaluation activities have addressed global project issues such as: Is the distribution environment increasing the accessibility of high quality course materials and flexibility of delivery for users in peripheral regions of Europe? Is the production environment improving the quality and efficiency of co-authoring processes for course producers and training providers? Is the learning environment really enhancing the quality of instruction for learners? and finally are the Virtual Campus and the demonstrators usable and user-friendly for each of its distinct user groups?

Initial conclusions from the evaluation report are very encouraging. The first goal of the MODEM project to integrate peripheral regions in Europe by offering high quality microelectronics education to learners within those regions has been achieved through the development of the MODEM virtual campus. The second goal of making available tools to decrease the teaching and preparation load of academic staff has also been achieved, in particular, through the development of the "pick and mix authoring" environment. Finally with regard to the third goal, to enhance the quality of instruction in microelectronics by including new instructional techniques, it is more difficult to respond as the tertiary institutes involved in MODEM are only now starting to employ distance education and tele-learning. However, initial results are promising. In conclusion, the evaluation results, both formative and summative, demonstrate that the MODEM project has made significant progress towards the projects original objectives in two years. The next logical step is to validate the MODEM environment within the European microelectronics industrial sector.

5 Conclusions

The increasing level of specialisation in the high tech sector generally and the microelectronics sector in particular has resulted in a shortage of human resources. In fact, skilled human capital is now recognised as the limiting factor to growth in the industry, both nationally and at European level. Staff in companies and universities require access to specialised training in order to maintain the relevance of their technical skills throughout their working life. Telematics technologies have the potential to transform the way education and training is organised and delivered and thereby provide a flexible solution to this problem. The MODEM project represents a first European action to explore and evaluate the use of telematics tools for customised, on-demand, flexible training of students, teachers and industry personnel. At present the MODEM software environment has been developed and is under evaluation. Initial results are very encouraging. Because of its generic nature, the MODEM environment is also directly relevant to education and training needs in other industrial sectors.

6 Acknowledgements

MODEM is partially funded by the European Union Telematics Application Programme (TAP) within EU Directorate General DGXIII. I would like to acknowledge all the project partners for their contributions to this paper. The advice from Mr. G. Weets and Mr. F. Sestini of the TAP Education and Training Sector Team in the ongoing development of the MODEM environment is also gratefully acknowledged.

7 References

[1] "Industry Forum - People - The Limiting Factor to Growth", SEMI, USA

[2] G. Crean and M. O'Sullivan. "Multimedia Optimisation and Demonstration for Education in Microelectronics (MODEM): A New European Microelectronics Telematics Based Educational Initiative", Proc. IEEE MSE '97, 127-128, 1997.

[3] F. Pécheux, Y. Herve, H. Marchal, N. Hertel, J.P. Stoquert and R. Stuck. "Learning by Virtual Doing: Protocol Simulators for Surface Analysis in Microelectronics", Proc. IEEE MSE '97, 30-31, 1997.

[4] P. van Rosmalen, J. Hensgens and B. Hahn. "Simulation-based Tele Learning" (in press), Int. J. Continuing Engineering Education and Lifelong Learning, Special Issue on: Multimedia and New Technologies in Education and Training.

[5] C. Viéville. "An Asynchronous Collaborative Learning System on the Web" (in press), The Electronic University by Springer-Verlag as part of the Computer Supported Co-operative Work series.

DISTANCE MEASUREMENT OF SUB-MICRON MOS DEVICES VIA INTERNET

Gregory LAPRADE, Emanuel MARQUES, CHEN XI, Etienne SICARD
INSA, Department of Electrical & Computer Engineering
Av de Rangueil 31077 TOULOUSE Cedex 4 - France
Fax: +33 561 55 98 00 -
e-mail: etienne@dge.insa-tlse.fr
web page : http ://www.insa-tlse.fr/~etienne

1 Abstract

The project concerning the internet measurement of sub-micron devices is undergoing at INSA Toulouse, with the objective to propose a permanent access to high-precision static current/voltage characteristics of sub-micron devices. A set of n-channel and p-channel devices, with short and long channels are already available in 0.7µm technology. A new set of devices fabricated in 0.25µm CMOS technology will also be connected and available in October 98.

2 Principle

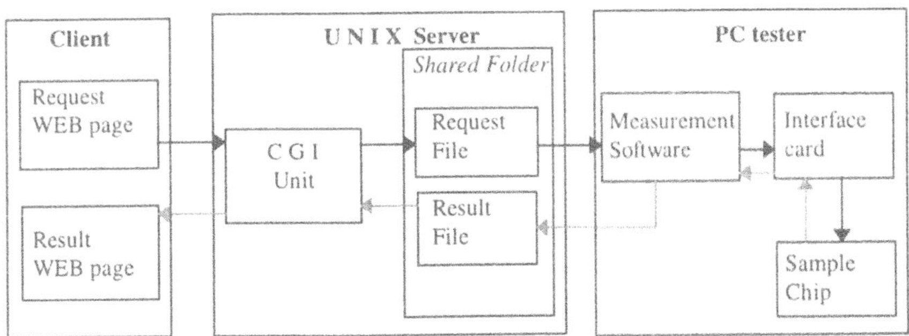

Figure 1 : Data flow of the Internet measurement principle

From figure 1, it can be seen that the Internet measurement system is divided into 3 parts : Client, Unix Server and PC tester.

T.J. Mouthaan and C. Salm (eds.), Microelectronics Education, 265-268.

Client : The clients can access our test chip directly by the HTML web page. Firstly, they set all the parameters needing for the test, such as the MOS type, size, the test range, and sampling number. A portion of the acquisition menu is reported in figure 2. After a click on the « send » button, the request is sent to the Unix server. Shortly after, the resulting curve is displayed on a new web page.

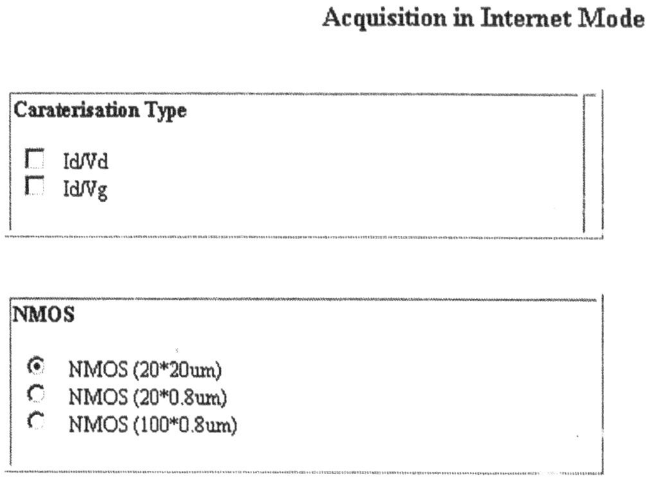

Acquisition in Internet Mode

Caraterisation Type

☐ Id/Vd
☐ Id/Vg

NMOS

◉ NMOS (20*20um)
○ NMOS (20*0.8um)
○ NMOS (100*0.8um)

Figure 2 :Client parameter setting page

Unix Server : The Unix server consists of two parts, the shared transmission folder and the common gate interface (CGI). The shared transmission folder is the common area for storing the request and the result files. The function of the CGI is to treat the request data and pass it to the shared transmission folder and on the other hand, it receives the result file and transfers it to the HTML format which will be finally shown on the client web page.

PC tester :

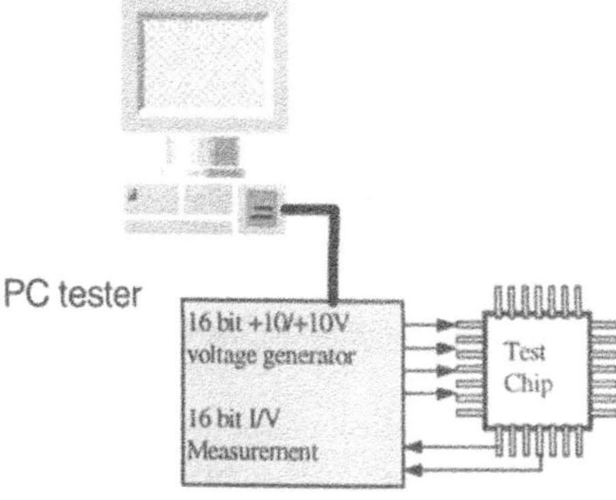

Figure 3 Architecture of the PC tester

The PC tester illustrated in fig. 3 is the main part of the measurement system. It consists of visual instrument platform to control the function of the card, the 16bit A/D card embedded on the ISA bus slot and the printed circuit board to hold the sampling chip. The PC tester scans the shared transmission folder of the Unix system to obtain a new request file, and then generates the input signals to stimulate the test chip according to the corresponding parameters. The result file is then transferred to the shared transmission folder.

MOS device measurement

The MOS devices connected to the measurement system have been fabricated in a CMOS dual-metal layer 0.7um technology. The chip called « FIBI » has been designed using the CAD tool « MICROWIND » developed in our institute [1]. Three MOS sizings are available directly on the extrenal pads with channel width and length respectively 20*20 µm, 20*0.8, 100*0 .8um. Details on the implementation of the MOS devices on the chip FIBI are reported in Figure 4. A total of six sizing are implemented on the MOS pattern. The pads appearing on the layout of Figure 4 are test pads for direct probing. The external pads, not appearing on the figure, are bonded to the package, routed on a printed circuit board and finally connected to the PC-compatible 16-bit measurement unit. The Id/Vg, Id/Vd characteristic curves can be measured via Internet, with voltage range between 0 and 7.0 V.

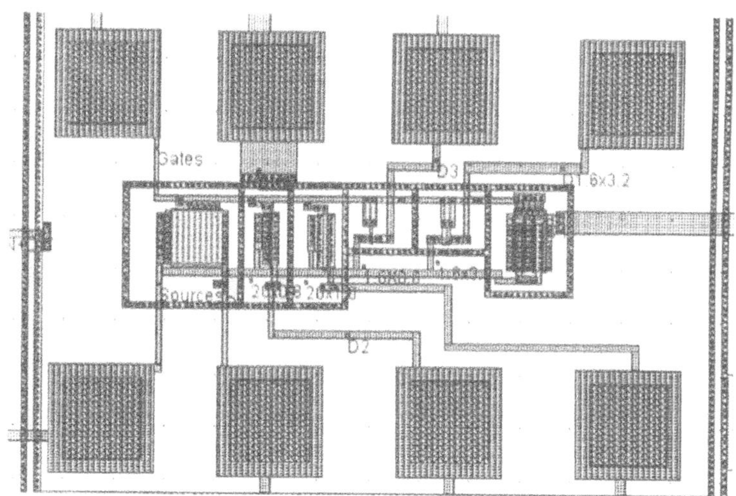

Figure 4 : The MOS devices connected to the measurement unit.

An interesting feature is the to compare the measurement result with the simulation based on the spice model 1 , 3 and Philips MOS model 9, interactively within the software Microwind, as shown in Figure 5. The level 1 is defined according to the ideal transistor theory based on the Schichman-Hodge equation available for long and large channels. Level 3 pursues a semi-empirical approach to model some short channel effects, but not fitting well with deep sub-micron technologies. Model 9 is a complex MOS-transistor model comprising a total of 60 parameters, for accurate simulation of logic and analog circuit for technologies down to 0.18µm channel length. After comparing the measurement with the simulation, the clients may adjust the parameters to watch its effects on the characteristics. In figure 5, the MOS model interface is shown, with the simulated and measured characteristics on the same screen, and a list of parameters that can be altered interactively.

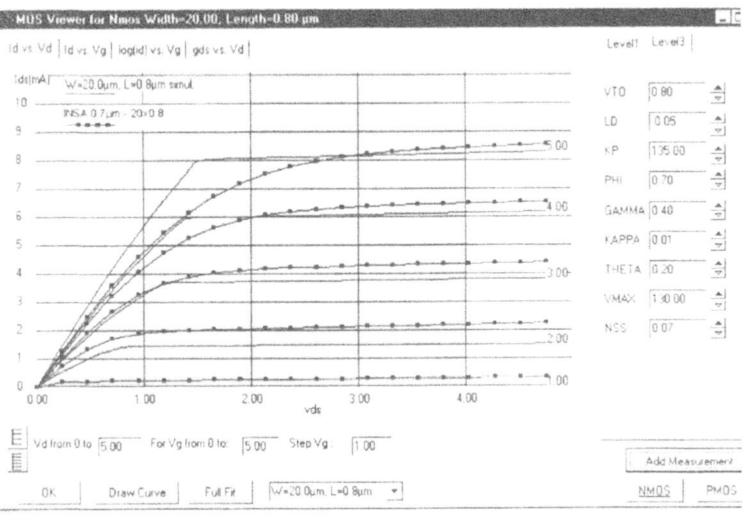

Figure 5 :Comparison of the measurements and the MOS model 3 in Microwind.

3 Conclusion

This project provides a convenient access to real measurement of sub-micron devices with no need to duplicate the costly measurement equipment. The Internet access to MOS measurements also proves to be very attractive to our students, and should be opened soon to students all over the world [2].

4. References

[1] E. Sicard, « Microwind, an introduction to microelectronics » User's Manual, Feb 98, INSA, Toulouse France

[2] Informations on site http ://www.insa-tlse.fr/~etienne/measure

A MULTIMEDIA LEARNING AND DESIGN SYSTEM FOR MICROPROCESSORS

R. MAYER, M. KOCH, D. TAVANGARIAN
University of Rostock, Department of Computer Science
Albert-Einstein-Str. 21, 18059 Rostock, Germany

1 Introduction

The increasing complexity of microprocessors as well as the modern principles of processor functionality and organization makes it more and more difficult to teach this topic with the classical methods. The new possibilities of multimedia tools are superior to teach and learn these complex informations. In reference to the experiences made with a multimedia learning system for VHDL [1,2] we decided to realize a new multimedia learning and design environment for microprocessor systems. Topics of special interest are the different processor components, their behavior, the various kinds of processor architectures and the programming of processors.

2 The Realization

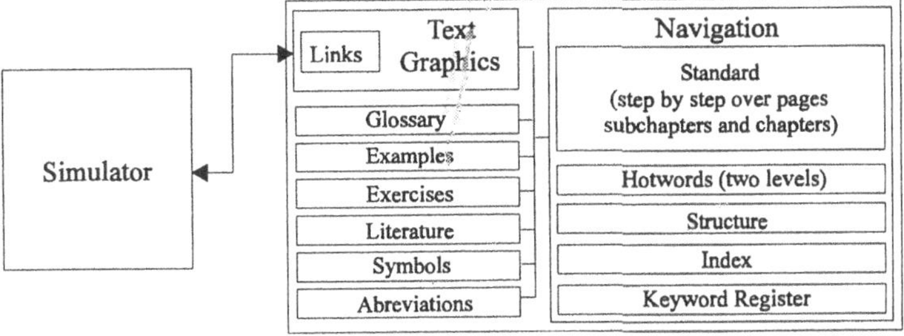

Figure 1. Hypertext based learning system

The developed tool consists of two interacting parts (Figure 1): A hypertext based learning system and a graphical simulation system for microprocessors. Following the conception and the functionality of this two systems are described.

2.1 THE MULTIMEDIA LEARNING SYSTEM

The hypertext based learning system consists of lessons which describe all basics of modern processor architectures. It also covers newer principles and features like pipelining, superscalar techniques, etc. The lessons are produced using an authoring system.

T.J. Mouthaan and C. Salm (eds.), Microelectronics Education, 269-272.

The whole instruction text is considered as a book, subdivided into chapters, subchapters and pages. The text of this book is extended by a lot of multimedia elements (animations etc.). To ease orientation for an user there are a lot of possibilities for navigation between the different parts of this book. The standard navigation is to leaf through the book page by page. Additional there is the possibility to leaf through subchapters and chapters and to use a two level hierarchy of hyperlinks (hotwords), an index register or a special keyword register. So the user is able to search for specific expressions and subjects. A history tool may be used to trace back a number of pages, the user visited before. After all there is the possibility to jump to additional parts of the book like the glossary, examples, exercises, literature etc. (Figure 1).

The user has two principle ways to read the book (Figure 2).

- In bottom-up-manner first all chapters describing the components of microprocessors may be read to learn all the details of their construction and behavior. Using this knowledge, then the chapter about the generation of a model processor (Figure 2 B) can be read and the microprocessor simulator can be used to serve the exercises of this chapter.

- In top-down-manner first the chapter about the model processor may be read and the other chapters can be used to look up to special points of interest while reading the model processor chapter (Figure 2 A).

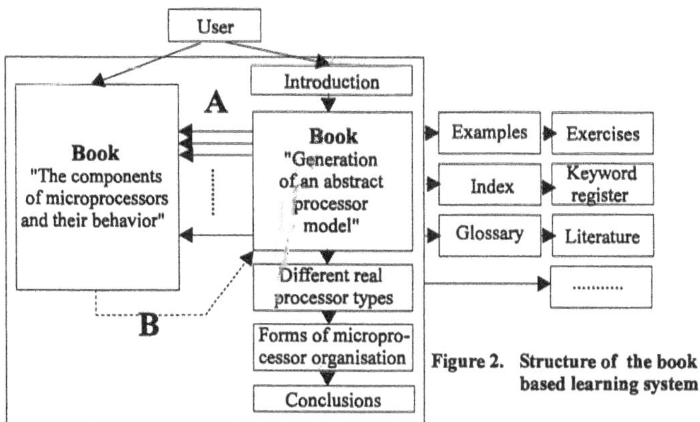

Figure 2. Structure of the book based learning system

Of course the user can use a combination of this ways, too. The individual way depends on the former knowledge and the habits of the user. To realize this individual way of learning, the main book is subdivided in two book parts which can call each other „The components of microprocessors and their behavior“ and „Generation of an abstract processor model“. Some other chapters are separated, because their content is independent of the main processor description. These are chapters about really existing microprocessors and microprocessor systems etc.

2.2 THE SIMULATION SYSTEM

The second part is an universal, graphical simulation system for microprocessors. It models the structure and the actions of real or theoretical existing microprocessor models. Finally it will be able to aid the user to design an own, new processor concept using a pool of components which is offered for the assembly of processors.

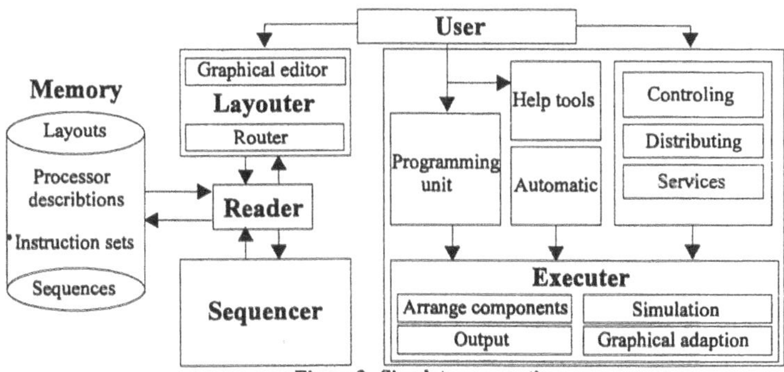

Figure 3. Simulator conception

Figure 3 shows the main elements of the simulation system. For the description of a microprocessor two main data files are needed. One contains a list of components of the processor. It also contains a list of the connections between the different components. The other file describes the list of the instructions of the processor. These instructions are subdivided into atomic steps. Before the first simulation of a new processor can be started, there is an interim, where a compiler produces two more data files. The first file is produced by the sequencer which determines the order of the processor components within the execution of a simulation. The second file is produced by the layouter which produces an interactive useable graphical representation of the components and their connections (Figure 4).

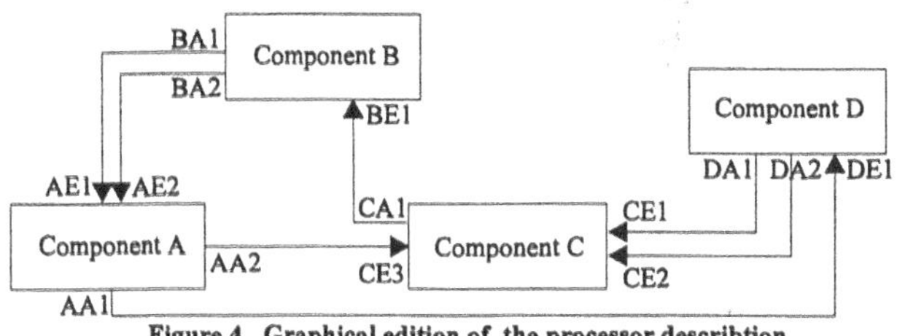

Figure 4. Graphical edition of the processor describtion

The sequences and the layout are saved as data files. Now the user is able to start the simulation by giving the simulator an assembler program. To write an assembler program the user gets support from a programming unit (editor, assembler etc.) and a help tool. Beside this the user has different possibilities to control the simulator, to choose the graphical appearance of the simulation, to decide the length of the simulation steps, to see details of the components, etc. Finally the executer uses the instruction set to interpret the assembler program. It can load the list of components into the processor data structure, controls the graphical representation of the processor, knows the order of the components for the simulation and gets the user commands. The algorithm of a simulation is illustrated in Figure 5. It is very simple because the simulator is produced by an object oriented programming language. For every processor component type there exists an own class. Every class contains the data which describes the state and the value of its component and all functions which describe the behavior of the component. At last the function which realizes the whole reaction of a component during a simulation has the same name for every component type. So the master program of the executer only has to call this function for every component.

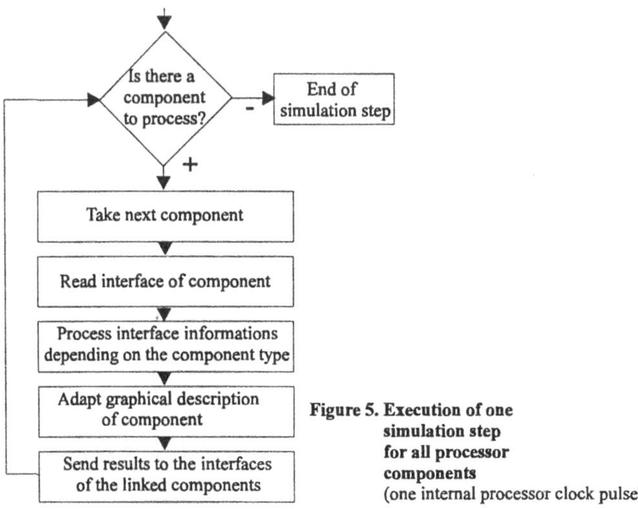

Figure 5. Execution of one simulation step for all processor components (one internal processor clock pulse)

Then each component „knows" what to do because of the functions of its own class. So the executer only has to fill the data structure of a class named „processor" with the values it found in the processor description files and to start the algorithm to realize the simulation. The class „processor" is only the master class of all component classes which the simulator knows. This is the component pool. Every processor, which can be simulated consists of a discrete partial amount of components out of this pool. So it is possible to create a new processor model because of a new combination of components out of the pool.

3 Conclusions

We have shown the structure for a learning and simulation system for microprocessors. Both systems are under development, prototypes have been implemented to validate the concepts and considerations.

4 References

[1] H. Dicken, M. Koch, D. Tavangarian: „The Hypermedia VHDL Learning System-Description and First Experiences", 1 st „European Workshop on Microelectronic Education", Grenoble, France, Feb 5-6, 1996

[2] D. Tavangarian, M. Koch: „VHDL!start - HTML-basierendes, multimediales Lernsystem Hypermedia VHDL", CD, ELRAD, Heise Verlag 1996

MULTIMEDIA TRAINING ON INTEGRATED CIRCUITS DESIGN WITH LOW COST HARDWARE

R.CASANUEVA, F.J. AZCONDO, M. MARTINEZ, S. BRACHO
Microelectronic Engineering Group. Dept. TEISA
ETSII y T. University of Cantabria
Av. de los Castros s/n 39005 Santander. SPAIN

1 Abstract

This paper presents a multimedia educational software for personal computers, oriented to the training on design and simulation of microelectronic circuits, where it has been tried to reproduce the actual design frameworks. The multimedia application is a tutorial that covers three widespread aspects of the design of ICs in the microelectronic design context.

2 Introduction

The education on microelectronics in the University of Cantabria has a solid background [1] that makes possible to undertake new teaching methodologies that improve the efficiency and reduce cost of the training tasks. In this way it has been launched an initiative to move part of the educational and training effort from the expensive UNIX workstation environment to personal computers by designing a multimedia software that contains an explanation, for self study, of the design-flow and simulation of digital integrated circuits based on the CADENCE design framework for the design with standard cells [2], Synopsys for simulation and synthesis [3] and Altera for implementations in FPGAs [4]. Thus, covering three significant digital circuit design techniques.

This work is being supported as a research project by the Spanish Ministry of Education and Science CICYT: TIC 95-0837-C02

3 Targets of the software

The software multimedia tries to collect the experience of the senior designers after several years participating in the ESPRIT programmes; Eurochip and Europractice. The resulting application is highly interactive and the students can easily recognize in the application the actual design environments, where the design activities will be carried out after the learning process. The educational software has been developed with Multimedia Toolbook, v4.0 [5] that allows the integration of graphic, sound and video resources

T.J. Mouthaan and C. Salm (eds.), Microelectronics Education, 273-276.

through an object oriented programming technique [6], where each object responds to an action defined in its associated script. The navigation system is based on a double way to access the information either directly or via a menu.
The general targets of these two activities are the following:
1) To spread the training on microelectronics to students that can not work in our laboratory for a long period; such us students from abroad, which is specially important to keep a collaboration with Iberoamerica and so do with students with not fully compatible activities (i.e. students from other universities or industry staff)
2) To achieve better efficiency from the laboratory facilities. Since the multimedia application runs into PCs, a first training stage can be developed in a low cost hardware outside the workstation network and is also more accessible to small and medium companies (SMEs).

4 Contents of the application

The software offers a clear information about three design frameworks: CADENCE, Synopsys and Max Plus II, that are available for the students of the different programmes with subjects in microelectronics. Additional chapters have also been included in order to solve the initial difficulties at the workstation, the Unix system and X-Windows. In the first stage of the navigation the student finds a general overview of the software that includes a video presentation of the faculty and laboratories, the targets of the tutorial, the multimedia resources and a small sub-program with the navigation instructions.
It has been followed some multimedia criteria to hold the attention of the student such as the colours and intensity of the screen backgrounds, the sound that confirms the mouse action and the music of the automatic tutorials.

4.1 CADENCE TUTORIAL

This first tutorial, about the CADENCE Design Kit, is focused on acquiring the necessary skills to design circuits based on standard cells as well as user designed cells. The tutorial puts special emphasis on the use of the Command Interpreter Window that gives access to the CADENCE tools by either choosing the menu options or writing commands.
The training method follows the design steps divided into six blocks plus an introduction, as is shown in fig.1 (left). Since Verilog-XL is the simulation tool whereby the students verify the design, the tutorial includes a chapter about the stimulus generation and simulation sequence with this tool. The tutorial gives also information about the place & route techniques and post-layout simulation. Within the Verilog framework, it is also included an overview of the use of Verifault, for fault simulation, and Veritime, for timing analysis.

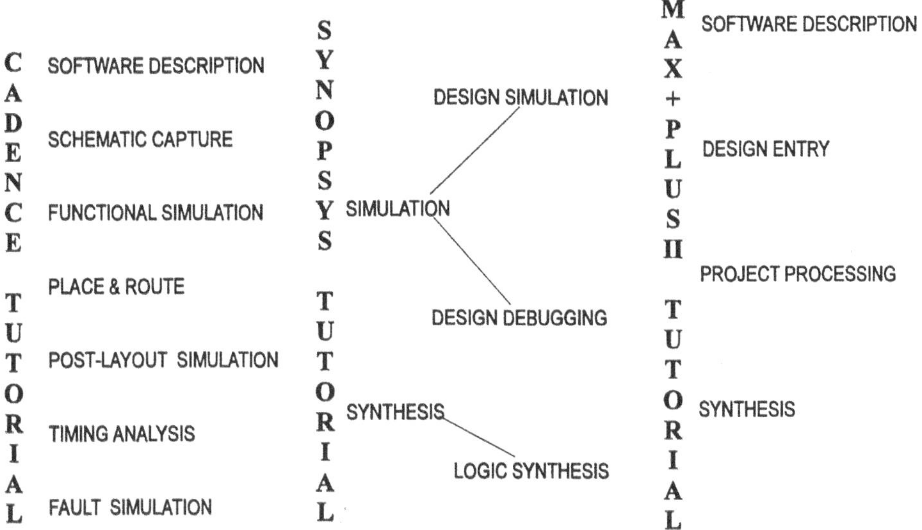

Figure 1. Chapters organisation in the multimedia software

4.2 SYNOPSYS TUTORIAL

The use of the tool Synopsys is described considering that the design entry is the VHDL description of a circuit. From this point, the student learns how to translate the design into a system at gate level (synthesis process) consistent with the restrictions given by the designer. The tutorial is divided in two sections: verification and synthesis according to the scheme of fig. 1 (middle)

4.3 MAX PLUS II TUTORIAL

This tutorial faces the explanation of the MAX PLUS system for the design with FPGAs from two points of view. The first sub-tutorial gives a general description of the design sequence and the second the student is guided through a practical design that consists of a circuit that generates signals whose frequency and pulsewidth are controlable.

5 Navigation through the tutorial

The contents of the multimedia software have been organized following the schemes and sub-schemes given at the starting level of the program, thus the students have always in mind the progress of their knowledge, the relation of a chapter with the whole information and may also compare the design stages of the technologies presented in this educational tool. A particular colour has been assigned to the text of each sub-tutorial, using white colour for the words that give access either to deeper information or to screens that represent the design environment, known as "hotwords".
Special care has been taken in making reversible most of the steps carried out by the user (see fig. 2), offering the possibility of returning to the main menu and exit the

application, with the exception of the most detailed information that may be found behind the hotwords.
Taking advantage of the multimedia capabilities it has been developed some automatic tutorials that consist of lessons about a detailed aspect of the use of the design framework that the student receives by mere observation of the chapter that includes animation and music for easier memorizing.
Navigation is performed by activating the intuitive icons available in the application screens that allow the user to exit, move to the next and previous page, return to the upper information level, return to the main menu and access to the automatic tutorial.

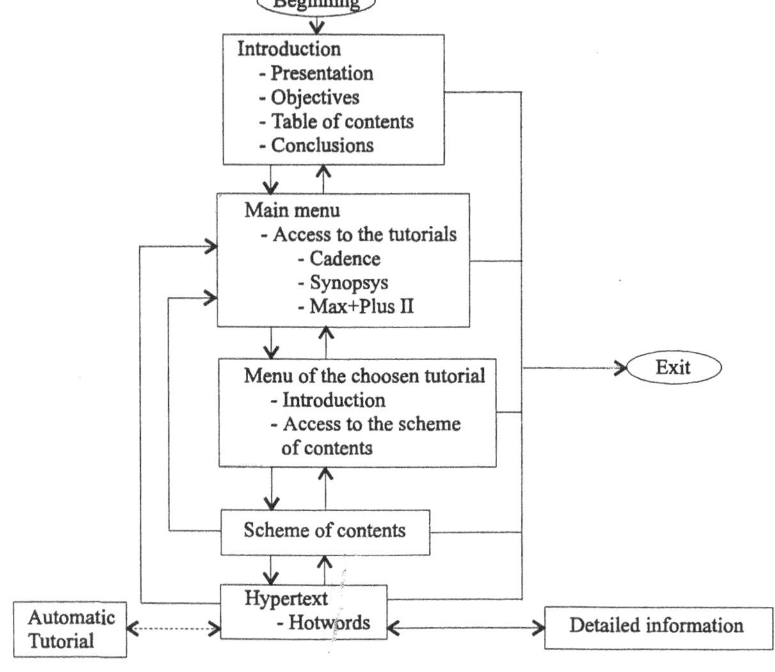

Figure 2 Navigation flowchart of the multimedia software

6 References

[1] S. Bracho, E. Villar, M. Martínez, P. Sánchez, M.A. Allende. VLSI Design training and education in the University of Cantabria within EUROCHIP. IEE Proc. G. Vol. 139, No.2 April 1992, pgs. 234-240
[2] "ES2 Cadence Design Kit". User Guide. 1995.
[3] Synopsys VHDL System Simulator Tutorial v 3.0b
[4] Altera Max + Plus II v 8.1 1997 Reference manuals
[5] "Using Toolbook". Reference Manual. Asymetrix Corp, 1994.
[6] "Openscript". Reference Manual. Asymetrix Corp, 1994.

Session F
DESIGN INNOVATIONS

THE DUTCH PICO-PROJECT ON EDUCATION OF MICROELECTRONICS

JAN DAVIDSE, professor emeritus,
Delft University of Technology, Department of Electrical Engineering, Mekelweg 4, 2628 CD, Delft, Netherlands.

Abstract

Pico is a successful project for the improvement of the teaching of the design of microelectronic systems and circuits at universities and engineering schools in the Netherlands. The paper describes the infrastructure that has been created for the fruitful cooperation of universities and engineering schools, along with the various actions taken in the framework of the project.

1 The need for microelectronics education

Microelectronics is an important field of activity in the Netherlands. It encompasses chip design and chip production, with Philips as a dominant player, as well as application oriented activities and software design. Consequently there is a sizable labor market. There is a need for all kinds of system designers, software specialists, chip designers and technologists.

2 The educational system in the Netherlands

In the Netherlands, much like in many other countries, we have two levels in higher education. There are three Universities of Technology, situated in Delft (founded in 1842), Eindhoven (1956) and Twente (1964) along with 19 Institutions of Higher Technical Education with a Department of Electrical Engineering. Figures 1 and 2 show their locations in the country.
From the beginning of the microelectronics era, the Electrical Engineering Departments of the universities have deployed considerable activity in the field of microelectonics. Lectures on the various aspects of microelectronics (technology, circuitry, chip design, system design) started as early as in the mid-sixties. Also, after much discussion, in the early seventies technological facilities for the fabrication of chips and other microelectronic structures were established in all three universities. At that time, the costs involved in founding such facilities were moderate. Much equipment could be purchased at highly reduced prices from industries that were modernizing their equipment, mostly because they switched over to larger wafer sizes.
In the Institutions of Higher Technical Education, further to be referred to as engineering schools, the conditions were less favorable. The E.E. Departments of most of these schools are small and consequently their teaching staff is also small. Whether or not a new topic, such as microelectronics, is taken up depends to a large extent on the motivation of one or two teachers. Although in some schools the subject was taken up adequately, in general the situation was for a long time unsatisfactory. Also, the contacts with the E.E. departments of the universities were unstructured. Where they existed, they mostly found their

T.J. Mouthaan and C. Salm (eds.), Microelectronics Education, 279-282.

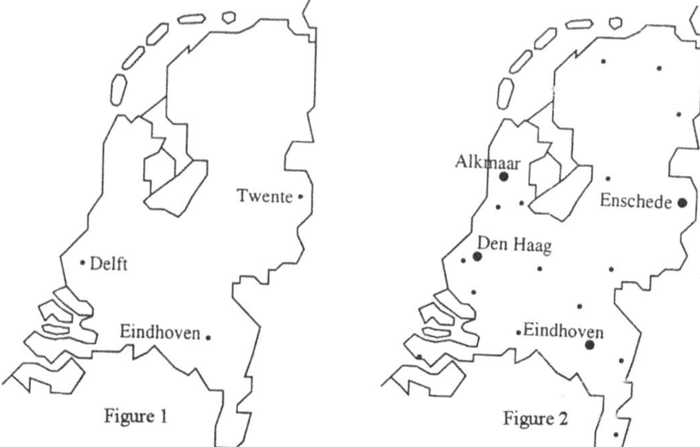

Figure 1

Figure 2

origin in contacts teachers kept alive with the labs where they had done their thesis work when being a student. Such contacts tend to fade out in the long run and henceforth are mostly restricted to younger teachers.

3 Recognition of the need for revitalization

In the early eighties the universities were heavily discussing plans for the near future. The existing fabrication facilities became outdated. Moreover, CAD was rapidly developing and for keeping up with the growth in this area, considerable investments in computers were necessary. A matter of much controversy was whether a common facility was tot be established at one of the universities, or three smaller ones. An even more important question was how to found the heavy investments that were to be done in whatever way of realization.

In 1985 the plans gained momentum, thanks to the pressure from the side of the industry on the government in view of the expansion of microelectronic activities and the shortage of adequately trained engineers. This pressure turned out to be effective. On short term funds became available for strengthening the educational infrastructure. It was decided that each university would have a facility for the manufacture of microstructures, each specializing in a certain branch of the field. Delft was to specialize on silicon structures, Eindhoven on 3-5 compounds and optoelectronics and Twente on sensor technology. Believe it or not, but the chosen formula has proven to be quite workable. The three institutions have grown into important centers of research and advanced education with large numbers of Ph. D. students. And their complementary facilities have very much favored cooperation.

Of great importance in the framework of the central theme of this conference is that, along with the funding of the research institutions, a considerable amount of money was allocated to start a project for the furtherance of education in microelectronic design in the undergraduate divisions of the universities and in the engineering schools. The sum available for this project amounted to Hfl 25 million (about M$ 12.5). An important condition was that the project should involve close cooperation between the E. E. Departments of the universities and the engineering schools. This would ensue that the engineering schools would rapidly share the knowledge already available in the universities. In order to organize the project a Steering Committee was formed consisting of three representatives of the universities, three representatives of the engineering schools and an independent chair-

man. As first chairman was appointed Mr. L. Tummers, a retired director of Philips Research Labs. The project was named PICO, Dutch acronym for 'Plan for stimulating the design of microelectronic circuits and systems at universities and engineering schools in the Netherlands'. Since structured contacts between the engineering schools and the universities, but also mutual contacts between the engineering schools, were virtually non-existent, the first thing to do was establishing an organizational structure that would promote the creation of mutual contacts and foster fruitful cooperation. This was done in the following way. Four clusters of engineering schools were formed according to a geographical pattern. In each cluster one school was designated as 'spearhead school'. In addition to each cluster one of the universities was assigned. Further, in each of the clusters a project coordinator was appointed, responsible for the coordination of the project activities within his cluster. And finally a project manager for the management of the whole project was appointed. Among his tasks is organizing regular meetings of the project coordinators. The committee consisting of the project coordinators and the project manager was assigned a twofold commitment. First the exchange of experiences and the furthering of overall coordination and second to serve as an advisory board for the purchase of hardware and software. The recommendations of the committee are reported to the Steering Committee through the project manager, who attends all meetings of the Steering Committee. Final decisions concerning purchases are made by the Steering Committee, that gathers further advices from an expert subcommittee.
Each cluster operates as a network to which all its institutions have access. The spearhead schools and the universities have extra facilities at their disposal that are available through the network.

4 Training of teachers

Having hardware and software facilities is, of course, indispensable for the teaching of microelectronics and particularly for training in microelectronic design. However, such facilities are of no avail without highly skilled teachers. In the universities much knowledge is available, because they are involved in research projects, often in an international embedding with the associated world-wide contacts through scientific conferences and exchange of reports, papers and Ph. D.-theses. However, the engineering schools are generally lacking these resources. Their teachers usually have full-time teaching commitments and the conditions for keeping up with the progress in the field are much less favorable. The Steering Committee was aware that providing opportunities for teachers to enhance their skills was of paramount importance. The Committee has organized a great many of courses for teachers. Most of these courses were structured and actually presented by the universities involved in the project. Also a considerable amount of teaching materials, such as lecture notes and textbooks, has been produced. These actions have brought forth an impressive quality improvement of the teaching of electronics in the engineering schools with the ensuing increase in the competence level of their alumni. Employers of young engineers are very content of their high professional level which enables them to be fully usable within a few months.

5 Development of the program

The project was started in 1986 with a foreseen duration of 5 years. Thanks to the great success of the project, in 1990 an additional budget of Hfl 8.33 million was made available for continuing the project up to the end of 1995. It was anticipated that thereafter the schools would be capable of maintaining their efforts on the basis of their regular budgets.

By careful management of the budget the Steering Committee was capable to end this second phase with a positive balance of over Hfl 1 million. This was accomplished mainly by creative negotiating with the suppliers of hardware and software to obtain sizable discounts on the purchase of their products. It stands to reason that trade and industry were eager to cooperate because of the amount of the consignments involved. Moreover, they were aware that familiarity of the students with particular products would assure them a preferential status when the former students would obtain responsibilities for purchases in their companies.

The government allowed the Steering Committee to use the remaining budget for continuing the project on a reduced scale up to mid-1998. This extension of the project was very welcome, because in the early nineties both the universities and the engineering schools were confronted with a very serious decrease of the enrollment of junior students. Regular financing depends on the number of students and henceforth the schools had to cope with declining budgets. Since the lion's share of the budget is spent on the cost of personnel, in fact part of the staff had to be dismissed. Not unnatural, the schools tried to restrict dismissals as much as possible at the cost of reduced expenditures on any other items. Although one million guilders for more than two years is not very much, it gave the Steering Committee the capability to keep the project on the road with the ensuing advantage of not disturbing the organizational infrastructure with its coordinators and the project manager. Also, the relatively cheap activities associated with the organization of courses could be maintained and even enhanced.

Unfortunately, the decrease of the enrollment of junior students is continuing and even worsening. This is a deleterious situation, not only because of the consequences for the universities and schools, but also because it sharpens the shortages in the labor market. Anyway these shortages will continue to exist for at least the coming five years.

The growing complexity of modern microelectronic products requires new and sophisticated software packages. If the schools are to keep up with the times they are bound to embrace the new design tools. This implies further investments in software and hardware. It may be true that, technically spoken, the present hardware is still in good order, generally it is not capable of supporting the new software. Since, due to the reduced enrollment, financing of new investments from the regular budgets is seriously jeopardized, in one way or another additional money has to be found.

In an intensive discussion with all the educational institutions, the Steering Committee has made an inventory of the present needs. It is interesting to note that the need indicated as most urgent is the continuation of updating courses and the production of educational material. Second comes new software and only third new hardware. However, the Steering Committee suspects that there is too much optimism concerning the continuing usefulness of available hardware.

Having made its inventory of needs, the Steering Committee has made up a new budget plan. It is assumed that the participating institutions continue to reserve part of their regular budget as a contribution to a new PICO-project, as has always been the case in the past. For additional money the Steering Committee is presently approaching the ministries of education and of economic affairs, along with representatives of the industrial world. Safekeeping of this highly successful project is, without any shade of doubt, important for providing the labor market with excellently trained engineers.

6 Conclusion

In conclusion, the success of PICO shows that it is possible to achieve at moderate cost nationwide improvement of education in microelectronics, by creating an organizational structure that, by its very nature, brings about coordinated action.

USING INDUSTRIAL ATE FOR TERTIARY ELECTRONICS EDUCATION: IS IT THE BEST SOLUTION?

S. DEMIDENKO, S. K. JHAJHARIA, T.-B. TAN, W.-Y. WONG
EC Department, Singapore Polytechnic
500 Dover Road, Singapore 139651

1 Abstract

Test technology is an important component of the curriculum in computer and electronics engineering. At the same time, as a subject it poses considerable challenge for both professors and students. The teaching concepts have to follow the evolution of test techniques, anticipate needs of the industry, as well as provide a relevant hands-on experience to tomorrow's entry level test specialists. The latest can not be achieved without the use of operational test systems to conduct practical sessions. The paper discusses applicability of industrial ATE in an educational environment.

2 Introduction

Traditionally, automated test equipment (ATE) manufacturers have been producing electronic test systems that meet the needs and demands of the industry and research organizations. There is however the other fairly big segment of ATE market that has not been addressed by the manufacturers - educational institutes and universities (our very conservative estimate of a number of technical universities and institutions having test technology in their educational programs shows that it surely exceeds several hundred).
Testing by its nature is an "applied" subject: it is hard to expect good results of education and training if the course does not include practical sessions with operational ATE. Students find it difficult to acknowledge and to correct their misconceptions and uncertainties unless it is done through hands-on sessions using a particular test system.
While many education institutions teach the subject, not all of them are in position of acquiring relevant test systems mainly due to the budget constrains. The other matter is availability of suitable test systems on the market. The conventional way used by universities is to try to get the best available industrial ATE. Is this the best approach? Do industrial test systems meet the needs of the technical educational establishments? From our point of view the answer is rather negative: due to a number of salient features and constraints inherent in the educational establishments, the direct use of manufacturing or engineering test systems can not be the best solution.
The aim of this paper is not to give a ready solution to the problem - we rather try to provoke a discussion on the matter. And we invite ATE manufacturers and users, academics and students, in short, all who are interested in the subject, to take part in it. In order to make the discussion more focused we limit ourselves by considering only IC testing in an undergraduate engineering course setting [1]. At the same time some of the issues presented in the paper are applicable to other kinds of testing as well as to other educational levels.

T.J. Mouthaan and C. Salm (eds.), Microelectronics Education, 283-286.

3 Curriculum

The first question to be addressed is the curriculum of the subject on test technology. Requirements to the tester are essentially determined by it - the tester has to be suitable to support teaching the topics presented in the curriculum. In this respect it is reasonable to determine a list of test techniques that are the most important for the industry, and the most widely taught. The result will give the set of functions expected from ATE that would suit the majority of educational institutions. Obviously, it is impossible to collect information on all courses on testing taught by higher universities world wide. However, as in any such analysis, it is important to use a fairly representative set of samples. This allows to make a grounded conclusion on the whole matter. To get the set we analyzed data from more than thousand different sources: electrical engineering academic programs on the World Wide Web (*http://www.ee.umr.edu/schools/ee_progtams_list.html*), Web-pages and prospectuses of particular universities, conference proceedings, technical literature, and others. In our research we limited ourselves by concentrating primarily on the courses on practical testing for undergraduate students (i.e., on training of "entry level test professionals" [2]).
Summarizing the results, we can conclude that in order to train the test professionals for productive work in today's and tomorrow's test technology, the curriculum, besides including the relevant basics (digital/analog electronics, IC design, IC manufacturing, data collection and analysis, computing and software engineering, networking, data communications, metrology and calibrations, etc. [1-7]) has to cover the following topics:

- ATE Architecture and Operation
- ATE Programming - Digital Test (Interpretation of DC and Timing Specifications of Digital ICs, Parametric (DC and AC) and Functional Tests)
- ATE Programming - Memory Test (SRAM/DRAM Organization and Addressing; Memory Test Patterns; Typical RAM Faults and Test Pattern Considerations)
- ATE Programming - Special Tests (IDDQ; Parallel and Serial Scan; BIST)
- ATE Programming - MCM Test
- ATE Programming - Analog Test (Op-Amp Specifications; Op-Amp Test Loop)
- ATE Programming - Mixed Signal Test (DAC/ADC Specifications and Test Circuits)
- Testing of Components of Special Types (RF, Low Power, 3D)

This list presents at the same time a general set of functions that are desired in the test system in order to make it an efficient teaching aid for education in test technology.

4 Test System

In contrast to the industry, an ideal tester for educational institutions is neither one with the highest fault coverage nor one with the highest throughput. In our case the most important among the requirements to the tester is to ensure that its functions, architecture, hardware/software tools along with technical parameters and characteristics correspond to the main goal of the educational establishment - to teach and to train future specialists in practical test technology [6]. Let us call this sort of considerations *the technical issues*.
The other important requirement to a tester for an educational institution is its cost effectiveness: basically, the budget limitations is the major constraint in acquisition and using hi-tech equipment for teaching and training purposes in the academic environment [8, 9]. Let us call considerations of this nature *the economic/management issues*.
In our case both the technical and economic considerations are closely related, and in our further discussion the separation between them will be rather relative. In fact, it is all too

easy to pay for tester technical features that are unnecessary or may never be used. Thus, optimizing technical specifications of the tester to the level that really satisfies the requirements of the educational process will at the same time keep the cost minimal - this is just an example of close interrelations between the two groups of issues.

Technical issues. First of all, the list of the tester technical capabilities is determined by the curriculum of the subject. As minimum the system has to support "core" types of tests, such as Parametric (AC, DC) and Functional Tests of digital and memory ICs, I_{DDQ}, Scan, Analog and Mixed-Signal Tests. The capability to cover as well other types of tests (such as those mentioned in the curriculum section) would be of additional advantage.

While being "wide" in terms of the test types, the system does not need to be "deep" in terms of its technical characteristics. Exceptionally high throughput, speed, resolution, accuracy are not critical for the systems used in academic institutions for educational/training purposes. From our experience, a tester with a single Parametric Measurement Unit, 2-3 programmable voltage supplies, nanosecond timing accuracy, and at least 32 independent bi-directional programmable pins with ECL, TTL and MOS signal levels is sufficient enough for teaching and training the entry level specialists in tests technology. With regards to memory ICs, the system has to be able to provide testing of SRAMs and DRAMs (with linear or multiplexing mode of addressing) of Mbit capacities.

Turning to tester software, it is necessary to stress that for the teaching/learning environment the importance of software comprehensiveness, operation simplicity and user friendliness shouldn't be underestimated. Possible syntaxical irregularities of the tester programming language could obstruct learning more than hands-on experience would aid it. Long and complicated program generation and debugging lengthen students' learning curve. Instead of focusing on test issues, students end up spending time and energy in their program generation and debugging. In this respect the software tools should be window-based with graphical programming capabilities. The system software should also provide interactive control and viewing for test parameters (e.g., driver and comparator voltages, power supply voltages, force and expect patterns). Availability of such debugging features as display of failing pins, waveforms and test vectors as well as editing and monitoring of digital activity are essential during interactive debugging. Programming "on-the-fly" is exceptionally useful in enhancing students' learning of device testing. An extensive library of test routines for standard cells and elements will also benefit the quality and shorten the learning process. Calibration, data collection and analysis tools would be useful in training and exposing the students to the kind of work in the test engineering field.

To facilitate "bridging" between two related subjects - IC Design and Testing (this is advocated by academics from different countries [2,5-8], the tester has to come with a software links to commonly used IC CAD environments (e.g., Cadence, Mentor Graphics).

In a typical class environment of twenty students, it would be impossible for trainees to have any sufficient program development time if a stand-alone tester with a single terminal is used. The solution is to incorporate a test system into a network environment where multiple programming stations are used for "off-line" test program development and debugging. In order to minimize the acquisition and operating expenses the "off-line" stations have to be of low-cost. Is it timely for ATE manufacturers to look into supplying programming and debugging tools that can be run on PCs or network computers?

Though the "off-line" stations allow students to simultaneously perform their test program generation, the students still have to queue up to use the tester to verify and run their programs. Given 2-3-hour laboratory session, each student has no more than 5-10 minutes of the tester time for running and debugging a program. This is obviously not sufficient to develop any serious skills in practical testing. Is it possible to have multiple independent test-heads in a tester that would allow students to run their programs simultaneously?

Economic Issues. By far, the main financial concern related to employing ATE in the academic environment is high initial acquisition cost. In fact it is much easier to the educational institution to cope with a relatively high operating cost rather than to justify even not too high expenditure for acquisition of a new ATE. We believe that by taking into account this circumstance and the nature of the usage of test systems in educational institutions the initial acquisition cost can be and has to be substantially reduced.

Besides above ideas of cutting the cost by optimizing the tester technical parameters to the level of the reasonable sufficiency, and by employing low-cost programming stations to realize a multi-user test generation/debugging, there are other perspective ways of reducing the acquisition cost, such as the *"pay-as-you-use"* scheme, and resource sharing.

Pay-as-you-use was initially introduced by the Hewlett-Packard in HP3070 board tester. We think the approach is worthy to be employed in the component tester area as well, especially for the systems meant for educational institutions. The essence of the approach is as follows. As discussed above, the ATE targeting educational institutions has to have a wide range of test capabilities and advanced features. However, not all of them are used permanently. The utilization period of a particular feature depends on the curriculum and the teaching schedule (in fact, some of the features are used for only 1-2 weeks in a semester). If ATE suppliers can charge a university an initial price for a basic system instead of the full amount, and recover the cost of the advanced features on a usage basis (for example, by means of special credit buttons as it is done in HP3070), then it will substantially facilitate an initial acquisition of the tester. Resource sharing is another way to cut the cost. The scheme can include several universities and small contract companies whose testing needs vary throughout a year. The ATE supplier will provide them with a basic system at a low cost while keeping a ready stock of hardware/software resources for the advanced features to be shared among the users as and when it is required.

Current advances in the communication technology give a basis to expect that the scheme where one company (let us call it a *test resources provider*) sells test services on-line to the consumers having special network terminals along with test heads and other minimally required equipment is no longer something unreal. Obviously, at this stage the approach is not suitable to testing in mass-production manufacturing where high throughput is the main criteria, however educational institutions and small manufacturers comprise the area where implementation of the sharing scheme will definitely be of benefit.

5 References

1. S. Jhajharia, H-S. Wang: Training Diploma Students on ATE-Related Module - ATS, 1995, pp. 184-188
2. A. B. Grubbs, Jr., G. Neland: Future Trends of Electronic Circuits with Implications for Entry Level Test Professionals - ITC, 1985, pp. 230-234
3. R. J. Feugate, Jr., S. M. McIntyre: Training Tomorrow's Test Engineers: Experiences in Teaching an Undergraduate Course in VLSI Testing - ITC, 1985, pp. 224-229
4. A. Miczo: What Do You Say When Writing a Text About Test? - ITC, 1985, pp. 236-238
5. E. J. McCluskey: Test Teaching - ITC, 1985, p.235
6. S. Mourad: Digital Testing Theory and Practice - ITC, 1988, pp. 205-206
7. R. Absher: Can Undergraduate Test Engineering Education be Faster, Better, Sooner? - ITC, 1991, p.1117
8. R. Ubar, J. Raik, P. Paomets, E. Ivask, G. Jervan, A. Markus: Low-Cost CAD System for Teaching Digital Test - 1st European Workshop on Microelectronics Education, 1996, Grenoble, France
9. T. Zimmer, F. Verdier, Y. Danto: IC Testing Course Performed in Industrial and Research Environment - 1st European Workshop on Microelectronics Education, 1996, Grenoble, France

TURBO TESTER: A CAD SYSTEM FOR TEACHING DIGITAL TEST

G. Jervan, A. Markus, P. Paomets, J. Raik, R. Ubar
Department of Computer Engineering, Tallinn Technical University, Raja 15, EE0026 Tallinn, Estonia. E-mail: jaan@pld.ttu.ee

1 Introduction

Traditional VLSI CAD/CAT systems for workstations are both costly and unable to handle large numbers of students simultaneously in educational courses. During recent years, many different low-cost tools running on PCs have been developed to fill this gap. They include the major basic tools for IC design: schematic capture, layout editors, simulators and place-and-route tools. However, low-cost systems for solving a wide class of tasks from the test domain, especially for teaching purposes, are missing.

A VLSI test design system for teaching graduate and undergraduate courses in integrated circuit test, called Turbo Tester (TT), has been created at Tallinn Technical University, Estonia. In the following, information about the features of this system is provided. Possibilities of using TT in educational courses is discussed. Finally, some future directions in the development of the system are considered.

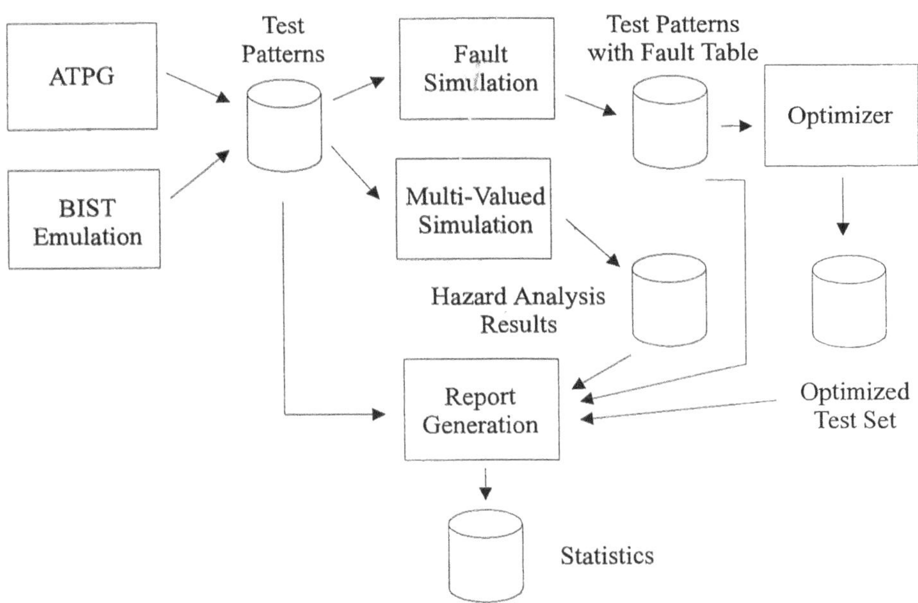

Figure 1. Data Flow of Turbo Tester

T.J. Mouthaan and C. Salm (eds.), Microelectronics Education, 287-290.

2 System Overview

2.1 TURBO TESTER TOOLS

Turbo Tester (TT) implements different methods for test generation (deterministic [1], genetic and random), simulation (static and multi-valued [2]), fault simulation (combinational and sequential circuits), test set optimization and testability analysis. Tools for Built-In Self-Test (BIST) simulation and quality analysis have been implemented, including structures for Built-In Logic Block Observer (BILBO) and Circular Self-Test Path (CSTP) approaches, which can be simulated and evaluated. Results of different tools and various statistics can be viewed by a special tool, called report generator. Figure 1 presents the data flow of interactions between TT tools.

2.2 GRAPHICAL USER INTERFACE

Turbo Tester Graphical User Interface (GUI) is shown in Figure 2. The tools of TT are presented as a visualized data flow, where the user can select and configure programs and form sequences of different diagnostic processes, which can be executed in a batch. The process output is displayed on a shell window. Similarly to most of the contemporary CAD systems, TT has a dedicated shell window with a command prompt.

In addition, the GUI includes a powerful C-like scripting language, which provides for conditional execution of commands and other traditional programming constructs. This allows the user to create customized batch files that execute the tools until satisfactory criteria have been met.

2.3 DESIGN INTERFACE

Many of the commercially available and in-house test design systems have problems with the design interface. Turbo Tester has a powerful design interface from EDIF 2.0.0 netlist format. It supports both, combinational and sequential designs. TT can read the schematic entries of various contemporary VLSI CAD tools, e.g. Cadence, Synopsys, Mentor Graphics, Viewlogic, Compass, OrCAD, ASYL+ etc.

2.4 CIRCUIT REPRESENTATION

The novelty of Turbo Tester lies in the fact that all the tools in the system are based on representing digital circuits in the form of Structurally Synthesized Binary Decision Diagrams (SSBDD) [3, 4]. SSBDDs provide a uniform approach to solving a wide scale of test design tasks, based on a uniform model and a restricted set of standard procedures. Unlike traditional Binary Decision Diagrams (BDD) [5], SSBDDs support test design for gate-level structural faults. In addition, all the tools of Turbo Tester can be used with gate-level descriptions, which makes it possible to compare different model representations. Moreover, in TT more general decision diagram representations can be synthesized, which allow migration of methods developed for logical level also to higher (behavioral and register-transfer) levels, where tools for hierarchical test generation and simulation have already been implemented [6].

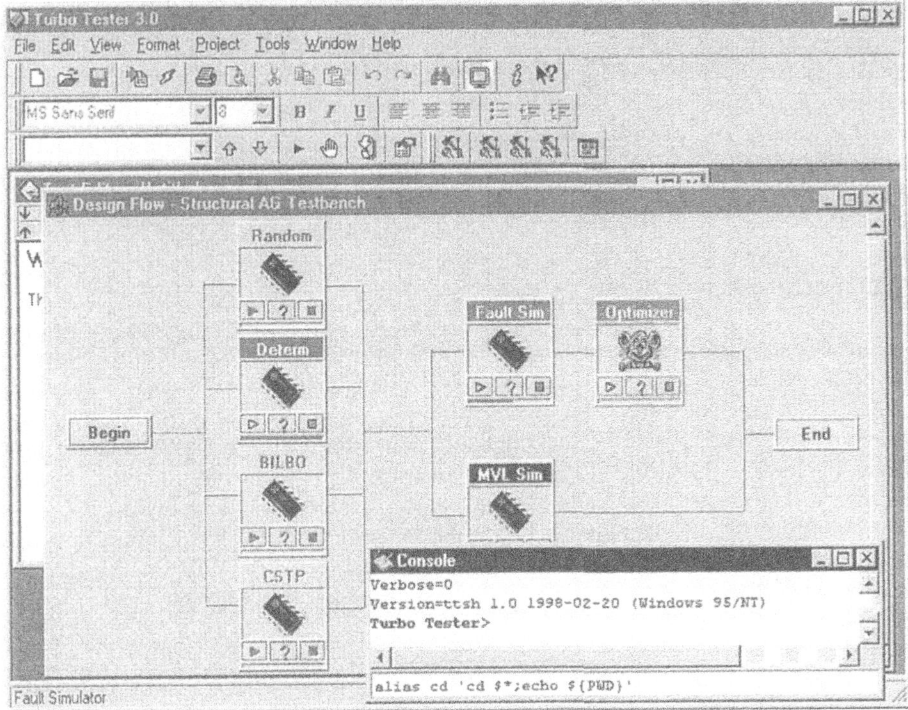

Figure 2. Turbo Tester Graphical User Interface

2.5 SYSTEM PORTABILITY

At present, Turbo Tester can be installed under MS Windows95/NT, Linux and Solaris 2.x operating systems.

2.6 USER DOCUMENTATION

Turbo Tester installation includes a comprehensive reference manual [7], where all the functions of the system are explained. The manual is designed in a style that is common to most of the CAD system documentations. The document complies partly with IEEE standard Std 1063-1987 for software user documentation.

3 Educational Courses

On the basis of Turbo Tester software, advanced laboratory courses have been developed whose aim is to teach and train students to integrate design and test, and to give them knowledge on how to create testable designs or designs with self-testing capabilities and how to obtain test patterns of better quality. In the courses the basics of test generation and fault simulation are explained. Students can compare different test generation (random, genetic and deterministic) and built-in self-test (BILBO and CSTP) approaches to examine the efficiency of methods with different fault models and with designs of different complexities.

The courses have received good credits from students of Michigan State University and Helsinki University of Technology, the software has been used in laboratory training in universities of Finland and Sweden, and it is under consideration to utilize Turbo Tester for teaching design for testability in other universities of Eastern and Western Europe.

4 Future Extensions

During recent years, a set of new tools that operate on higher design abstraction levels has been created in Design and Test Center of Tallinn Technical University. These tools include a design interface from register-transfer level VHDL, a hierarchical test generator [6] as well as tools for simulation and dependability assessment. In near future, the tools will be added to the Turbo Tester system as an extension, or a separate package will be created that would use some of the low level functions of TT.

In addition, we are planning to widen the range of fault models supported by the system, develop new tools for diagnostic analysis and create tutorials for assisting the students in studying the basics of digital test and diagnostics with the aid of TT.

5 Acknowledgements

Development of Turbo Tester has been partly funded by SYTIC CP96-170 project and Estonian Science Foundation grant G-1850.

6 References

[1] P. Goel, "An Implicit Enumeration Algorithm to Generate Tests for Combinational Logic Circuits", *IEEE Trans. Comp.*, vol. C-30, pp.215-222, Mar. 1981.

[2] R. Ubar, "Dynamic Analysis of Digital Circuits with 5-Valued Simulation", *Mixed Design of Integrated Circuits*, pp. 187-192, Kluwer Academic Publishers, 1998.

[3] R. Ubar, "Test Generation for Digital Circuits Using Alternative Graphs", *Proc. of Tallinn Technical University, Estonia*, No. 409, pp. 75-81 (in Russian), 1976.

[4] R. Ubar, "Test Synthesis with Alternative Graphs", *IEEE Design and Test of Computers*, Vol.13, No. 1, pp. 48-57, Spring 1996.

[5] S.B. Akers, "Binary Decision Diagrams", IEEE Trans. on Computers, Vol. 27, pp.509-516, 1978.

[6] M. Brik, G. Jervan, A. Markus, J. Raik, R. Ubar, "Hierarchical Test Generation for Digital Systems", *Mixed Design of Integrated Circuits*, pp. 131-136, Kluwer Academic Publishers, 1998.

[7] *Turbo Tester Reference Manual*, Version v3.0, Tallinn Technical University, Estonia, March 1998.

A PROJECT-ORIENTED UNDERGRADUATE COURSE IN MIXED ANALOG/DIGITAL ASIC DESIGN

L. HEBRARD, C. PETER and F. BRAUN
ULP - Pôle CNFM MIGREST - LEPSI
23, rue du Loess B.P. 20 - 67037 STRASBOURG - Cedex 2 (FRANCE)
http ://www-lepsi.in2p3.fr

1 Introduction

Microelectronic systems design for ASICs is still considered as a hard and very specialized topic by undergraduate students in electrical and electronics engineering (EEE). This feeling is justified because the mastery of ASICs design requires wide knowledge in fields as different as semiconductor physics, VLSI fabrication, circuit theory, signal theory,... Nowadays, the progress in technology have lead to more and more complex developments where logic and analog functions are integrated on a same chip. A future designer has to be prepared to these new challenges. It comes into its own that the task is not easy to fullfill. Many students can be put off if no effort is made by the education to promote the exciting world of microelectronics.
This analysis prompted a new introductory course in ASICs design for undergraduate students in EEE (Maîtrise EEA) at the University of Strasbourg (France). Microelectronics education cannot restrict to theoritical studies, but practical ICs design using professional CAD tools. Thus we challenged the students to carry out their own circuits up to the characterization of the final product.

2 Course contents

A single slope 4 bits Analog to Digital converter appeared to be a good choice that includes in a same project every aspects of mixed ASICs design. Here, analog and digital design aspects are approached, also allowing an introduction to analog-digital mixing bounded problems.
Figure 1 shows a block diagram of the well known single slope A/D converter [1][2]. It consists of two analog modules (a ramp generator and a comparator) and one logic module (a counter together with its control logic).
At the begining of the course, the specifications required for the converter are given to the students together with a document intended to guide them during all the design project. This document consists of two parts. The first one is an analysis of the project. It

T.J. Mouthaan and C. Salm (eds.), Microelectronics Education, 291-294.

is built up like a "large exercise" which leads the students to a first design set. At this point, the students are ready to start the practical conception of the circuit using CADENCE CAD tools. The second part of the document explains the students how to simulate, to draw a layout and to post-simulate the circuit up to the final post-mixed-simulation.

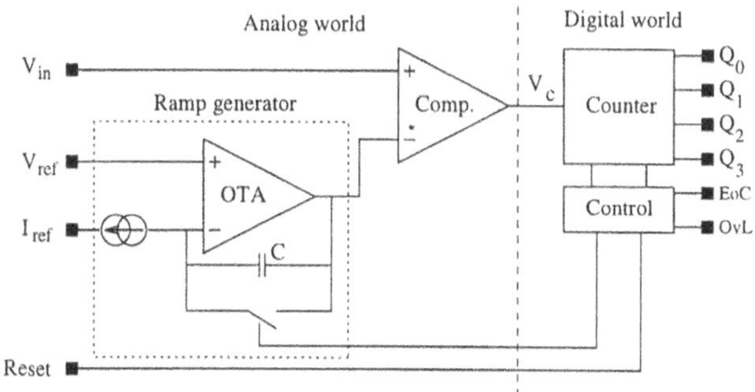

Figure 1 : Single slope ADC

Twenty hours of lecture are used to set up the theoritical study of the project. Most of the students attending to this course have a very poor knowledge about microelectronics. Therefore, about six hours are needed for presenting some basic aspects of microelectronics relevant for the project (transistor models, SPICE and VERILOG simulators, circuit layout, standard-cells design versus full-custom design, ASICs top-down design methodology) [3][4]. The questions asked in the "large exercise" [5] mentioned above are solved during the remaining fourteen hours using a top-down design methodology. We think that applying this methodology on an actual project is a great advantage over more conventional approaches in ASICs design teaching. It gives the students a deeper understanding of the proper design method.

3 Practical ASIC design using CADENCE

Fourty hours of CADENCE Design Framework II practicals are devoted to the conception of the converter. The project meets deadlines. So we have placed milestones to guide efficiently the students whilst designing. The main steps are summarized in table 1.

Functional block	**Hours**
OTA or Comparator *(SPICE simulation, layout, DRC-LVS, post-simulation)*	20h
Ramp generator	4h
Counter and control logic	9h
Final assembly *(Final layout, DRC-LVS, Mixed post-simulation, GDS2 file)*	7h

Table 1 : Practical design main steps

At the begining of the praticals, we provide the students with a document to guide them through the use of CADENCE. Then, we let them manage their own project, with our help if needed. When they fail to achieve a functional block in time, we provide them with it. Our first experience in this course showed that most of the students were very motivated and had spent many extra-time (from 30% to 100% of the initial fourty hours) on completing their circuit.

4 First results

The circuit is designed in standard 1.2µm CMOS technology and sent for manufacturing by means of the French Multi Project Chip service. Figure 2 shows the layout designed by a student and sent to fabrication recently. In the upper part, we recognize the logic module. The two modules on the bottom right are the comparator and the ramp generator. In order to visualize the ramp, a unit gain buffer has been added. This buffer and the pad ring are provided to the students.
During the design, we did so as to enable a separate test for each block (the ramp generator, the comparator and the counter). This shows the students that thinking of testability during the first design steps is a key point to be able to test the circuit successfully.

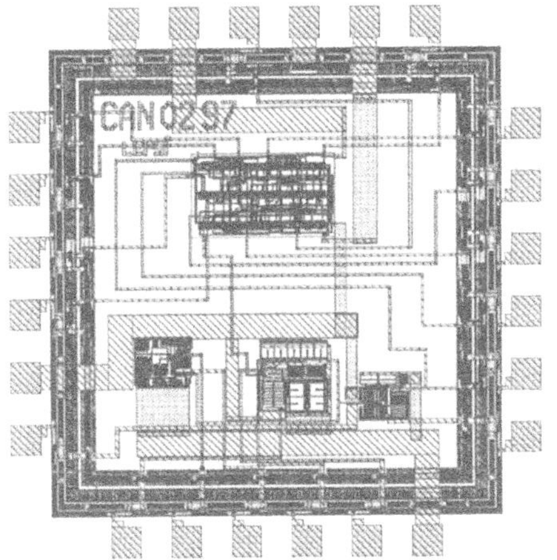

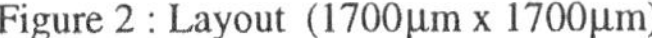
Figure 2 : Layout (1700µm x 1700µm)

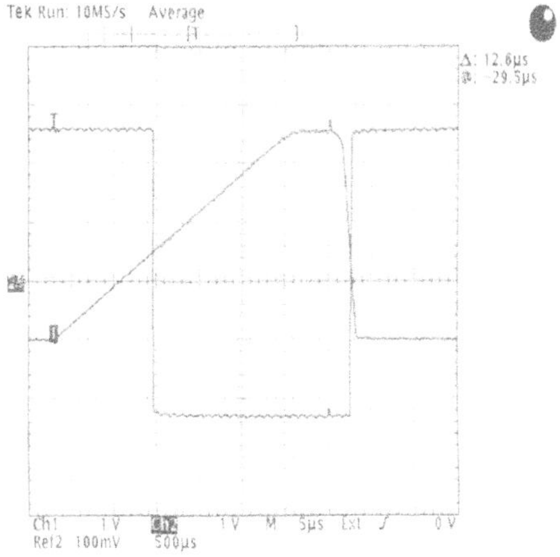

Figure 3 : Ramp generator and comparator output

5 Circuit characterization

The remaining four hours are spared for thoroughly characterizing the converter. The students successively test each module of it and end up testing the whole converter. For instance, figure 3 shows the output of the ramp generator (Vr) together with the output of the comparator (Vc). The negative edge on Vc ends the conversion (figure 3 and 4), while the positive edge resets the converter. During this reset, the speed at which the signal Vr returns to its initial voltage is limited by the slew-rate of the OTA. Thus, the slew-rate can be evaluated (around 2V/μs). By using similar test setups, the students successively measure the full power bandwidth, gain bandwidth product, offset of the OTA... The next step is the evaluation of the response time of the comparator (figure 5). Finally, they measure the transfer characteristic of the converter. An external logic signal also allows to disconnect the analog part from the digital part in order to enable a counter-only test. Of course, all these measurements are compared with the simulation results.

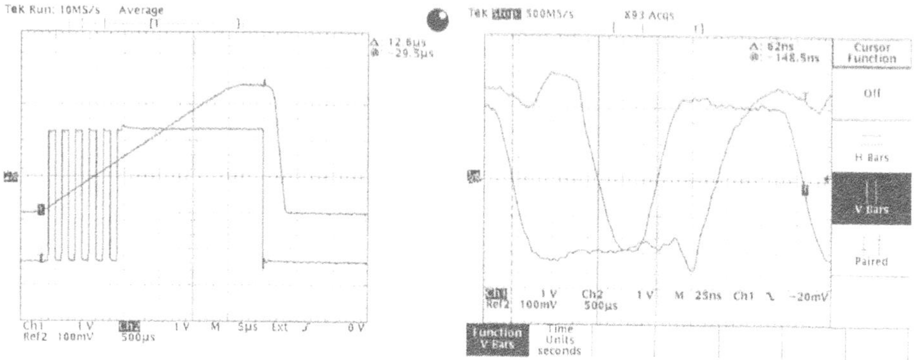

Figure 4 : Ramp generator and Q0 output

Figure 5 : Comparator response time ($\Delta t = 62ns$)

6 Conclusion

This new educational introductory module in ASICs design has been successfully acknowledged by the students. Most of them have devoted extra hours to finish their circuit before the manufacturing deadline. A key point to motivate them, and so to make the learning of microelectronics easier, was to work on a project they have to manage. The main features of modern ASICs design except high level synthesis have been tackled during the course. We think that this missing part is necessary even for an introductory course. So, we foresee to introduce it for the next academic year by the use of SYNERGY to synthesize the logic module of the converter, which will include a boundary scan path to introduce new basic aspects of testability.

7 References

1. R van de Plassche, ''Integrated A/D and D/A Converters'', *Kluwer Academic Publishers*, 1994
2. P. Allen, D. Holberg, ''CMOS Analog Circuit Design'', *Ronehart and Winston*, 1987
3. K. Laker, W. Sansen, ''Design of Analog ICs and Systems'', *McGraw-Hill*, 1994
4. N. Weste, K. Eshraghian, ''Principles of CMOS VLSI Design'', *Addison-Wesley Publishing Company*, 1993
5. L. Hebrard, C. Peter, ``Cours de CAO microélectronique en Maîtrise EEA'', Accessible on http://www-lepsi.in2p3.fr

PERSPECTIVES OF RECONFIGURABLE COMPUTING IN EDUCATION

J. BECKER, F.-M. RENNER, M. GLESNER
Darmstadt University of Technology
Institute of Microelectronic Systems
Karlstr. 15, D-64283 Darmstadt, Germany
e-mail: {becker, renner, glesner}@mes.tu-darmstadt.de

1 Abstract

The paper describes the emerging potential and flexible properties of reconfigurable computing in education of engineers and students, as well as of technicians of small and medium sized enterprises (SMEs). Therefore, an FPGA-course currently under development at Darmstadt University of Technology is outlined. The dedicated structure and contents cover an introduction into reconfigurable computing and its applications, cost/performance trade-offs (e.g. for small companies), fine-/coarse-grained and hybrid device architectures, overview on vendors and corresponding design tools and the structural programming process including examples from different areas.

2 Introduction

Configurable Computing demonstrates currently its potential of achieving high performance improvements for a wide range of applications like image processing [1], [2], [3] and compression [4], morphology [5], feature extraction [6], computational chemistry [7], object tracking [8], fuzzy controllers [9] among many others. The obtained performance results deliver an order of magnitude improvement over general purpose microprocessors. Due to this real speed-up potential a lot of researchers have built corresponding (re)configurable prototype systems for a large number of applications. The area is very promising and receives now more and more popularity, proved also by the first published article on this subject from the main stream periodical "Scientific American" [10]. FPGAs (Field Programmable Gate Arrays) consist of arrays of configurable logic blocks (CLBs) implementing the logical functions of gates. Both the logical functions as well as the interconnections between the corresponding blocks can be altered in seconds by downloading bitstreams to the chip. The programming of such "structurally programmable" devices can be repeated infinitely often, even partially and dynamically during run time. For more details on reconfigurable architectures, design tools and applications see [11], [12], [13]. Current computer science curricula do not create awareness, that hardware has become soft, nor, that hardware, structural and sequential software are alternatives to implement problems. Principles and applications of dynamically reconfigurable devices as a basis of the new computing paradigm of *structural programming* should be included in academic and industrial main courses to increase the potential of this novel emerging technology.

T.J. Mouthaan and C. Salm (eds.), Microelectronics Education, 295-298.

3 Structure and contents of the FPGA-course

The FPGA-course currently under development at Darmstadt University of Technology is addressed to different groups of persons ranging from students over researchers and educators to people from small and medium enterprises intending to get introduced to this technology.
Starting from an introduction to reconfigurable computing and the available technologies of field programmable logic in different views (depending on the learning person), a cost/performance trade-off is described. This enables the learner – a student to get an idea of the "what is possible and what is realizable"-problem, a learner interested in this technology to get an idea what he has to spent to do reconfigurable computing at home and an industry person (e.g. a salesperson) to calculate the possibility to use this technology for new products. The next step includes a description of what kind of tools are available, how is the situation on the market, who are the prime vendors, etc. This section gives a broad overview on the possibilities of reconfigurable computing and the corresponding design environments. Therefore, the structures and key differences between the various realizations (e.g. architectures) of field programmable devices are explained, including a number of the major components.

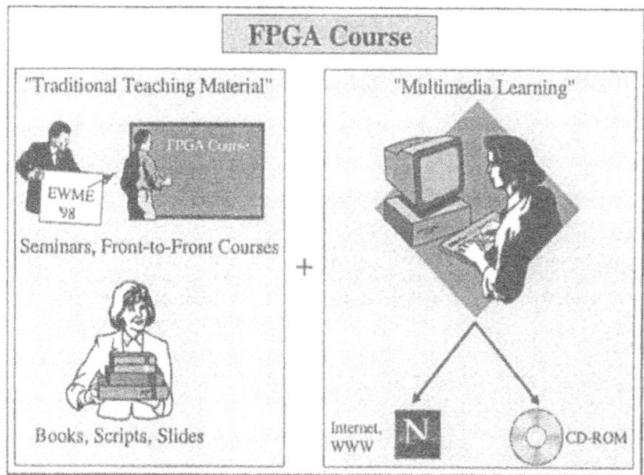

Figure 1: Integration of traditional and multimedia teaching techniques within the FPGA-course

The following section includes a detailed guide to the programming techniques and available tools of reconfigurable devices. Starting with simple examples the course candidate is guided through the designflow considering a list of requirements and the system specification through implementation and verification resulting in the finally downloaded bitstream for a reconfigurable device. The theoretical part of the FPGA Workshop is concluded with various examples from different application areas like automotive or mechatronic applications (e.g. Finite-State-Machines, Controllers, etc.). The candidate is able to select the amount of work he has to do to complete this example. A candidate who wants a detailed example is able to work on this problem largely on his own, whereas someone needing more help is more guided through the design process. In the final version of the FPGA-course the candidate will be able to communicate over the internet with a tutor to ask questions on certain implementation problems, to discuss possible solutions and to get new information and updates.

While most of these course sections are theoretical, e.g. the candidate has to study an amount of data on paper, or electronically on the web/CD-ROMs, a key aspect of this FPGA-course is the possibility to program a real device over the Internet and to see the results on the screen (see Figure 1). This enables the candidate directly to see whether his implementation is working on a real device or not. In addition this enables the instructor to realize the problems candidates face and if these problems refer to a lack of information within the FPGA-course.

4 Implementation concept of the FPGA-course

The introduced FPGA-course includes traditional education concepts such as front-to-front seminars hosted by an experienced designer using slides as well as student sessions. The candidate can choose to attend a conventional course at Darmstadt University of Technology, and/or to study at home using Computer Based Training (CBT) techniques such as a CD-ROM, providing a fast access to all information of the FPGA-course.

The key aspect of our FPGA-course are the practical sessions and the realistic use of reconfigurable hardware. The participants should be able to put the theory into practice by solving experimental tasks with real hardware devices.

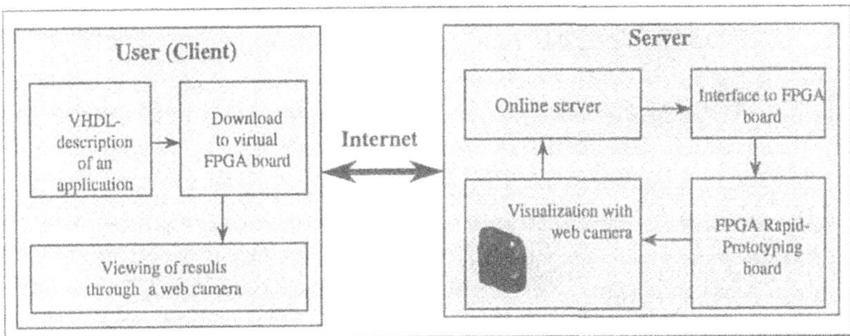

Figure 2: System architecture of realistic structural programming concept

Issues and applications from many different areas can be used to focus the course depending on the candidate's fields of interest. The candidate may begin with an easy traffic light controller to be developed and simulated. Next, the structural code can be used to program a reconfigurable PCI-bus based extension board (H.O.T. Works [14], or ProTest [15]) connected to the user via the Internet (see Figure 2). The results of the traffic light controller example would be the changing of LEDs on the FPGA board, visible through a web camera. More experienced users are able to implement complex algorithms (like signal processing and image processing algorithms) using similar implementation techniques and success control mechanisms. If the candidate has problems or questions, communication with a tutor via the internet can be established. In a first realization the candidate is able to discuss questions with a chat based mechanism including a whiteboard to visualize figures and implementation details. In addition this can be used to give new exercises and updated information to the candidate. The tutor sends the exercises to the candidates and gives details/hints for the solution. The candidates are able to directly address the tutor for further understanding purposes. In a future course implementation the conversation between learner and tutor will be realized with an audio/video connection using MPEG4 techniques.

5 Conclusions

The paper sketched the structure and contents of a theoretical/practical course in reconfigurable computing, developed for different severity levels (students, technicians, engineers). The authors believe that future computing machines and digital systems 10 years from now will include a strong mix of sequential programmable hardware and structural programmable reconfigurable devices, because the corresponding structural programming environments can be mixed with traditional programming tools. More course-grained reconfigurable device architectures, which are able to implement large and complex applications, may be integrated transparently as accelerators onto microprocessor programming environments, or used as alternatives for pure ASIC solutions [16].

6 References

[1] P. Athanas, A. Abbot: Real-Time Image Processing on a Custom Computing Platform, IEEE Computer, vol. 28, no. 2, Feb. 1995.

[2] R. W. Hartenstein, J. Becker et al.: A Novel Machine Paradigm to Accelerate Scientific Computing; Special issue on Scientific Computing of Computer Science and Informatics Journal, Computer Society of India, 1996.

[3] D. Ross, O. Vellacott, M Turner: An FPGA-based Hardware Accelerator for Image Processing; Proc. of 3rd Int'l. Workshop on Field-Programmable Logic and Applications, Oxford, Sept. 7-10, 1993.

[4] R. W. Hartenstein, J. Becker et al.: A Reconfigurable Machine for Applications in Image and Video Compression; Europ. Symp. on Advanced Networks and Services, Conf. on Compression Technologies & Standards for Image & Video Compression, Amsterdam, The Netherlands, March 20-24, 1995.

[5] T. H. Drayer, W. E. King, J. G. Tront, R. W. Conners: A MOdular Reprogrammable Real-Time Processing Hardware, MORPH; FCCM'95, IEEE Computer Society Press, Napa, CA, April 1995.

[6] A. L. Abbott, P. M. Athanas, L. Chen, R. L. Elliott: Finding Lines and Building Pyramids with Splash 2; IEEE Workshop on FPGAs for Custom Computing Machines, FCCM'94, IEEE Computer Society Press, Napa, CA, pp. 155-161, April 1994.

[7] J. Becker, R. W. Hartenstein, R. Kress, H. Reinig: High-Performance Computing Using a Reconfigurable Accelerator; Proc. of Workshop on High Performance Computing, Montreal, Canada, 1995.

[8] M. Shand: Flexible Image Acquisition Using Reconfigurable Hardware; FCCM'95, IEEE Computer Society Press, CA, April 1995.

[9] T. Hollstein, A. Kirschbaum. M. Glesner: A Prototyping Environment for Fuzzy Controllers; 7th Int'l. Workshop On Field Programmable Logic And Applications, FPL'97, London, UK, Sept.1-3, 1997, LNCS 1304, Springer Press, 1997.

[10] J. Villasenor, W. H. Mangione-Smith: Configurable Computing; Scientific American, June 1997.

[11] W. Moore, W. Luk: 5th Int'l. Workshop on Field-Programmable Logic (FPL'95), Oxford, UK, 1995.

[12] R.W. Hartenstein, M. Glesner: 6th Int'l. Workshop on Field-Programmable Logic: Smart Applications, New Paradigms and Compilers (FPL'96); Sept. 1996, Darmstadt, Germany.

[13] W. Luk, P.Y.K. Cheung, M. Glesner: 7th Int'l. Workshop on Field-Programmable Logic & Applications; Sept. 1997, London, UK.

[14] H.O.T. Works, Virtual Computer Corporation, http://www.vcc.com

[15] ProTest, Biel School of Engineering, http://www.isbiel.ch/I3S/e.html

[16] J. Becker, A Partitioning Compiler for Computers with Xputer-based Accelerators; University of Kaiserslautern, 1997.

The manufacturer's authorised representative in the EU is Springer Nature Customer Service Centre GmbH, Europaplatz 3, 69115 Heidelberg, Germany. If you have any concerns regarding our products, please contact ProductSafety@springernature.com

Printed and bound by CPI Group (UK) Ltd, Croydon, CR0 4YY

23/04/2026

02095593-0002